Applications of Polymer Spectroscopy

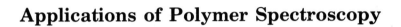

CONTRIBUTORS

G. E. ADAMS

F. J. BOERIO

BERNARD J. BULKIN

D. A. CHATFIELD

HUNG YU CHEN

M. M. COLEMAN

I. N. EINHORN

J. H. FUTRELL

ROBERT V. GEMMER

MORTON A. GOLUB

MICHAEL A. GRAYSON

R. J. GRITTER

F. D. HILEMAN

HARVEY H. HOEHN

B. L. JOESTEN

RICHARD A. KOMOROSKI

J. E. MCGRATH

LEO MANDELKERN

G. DAVID MENDENHALL

G. J. MOL

RICHARD A. NATHAN

T. E. NOWLIN

DERICK W. OVENALL

P. C. PAINTER

L. M. ROBESON

MARK L. ROSENBERG

DAVID K. ROYLANCE

W. O. STATTON

ROGER S. STRINGHAM

MADELINE S. TOY

K. J. VOORHEES

E. W. WISE

CLARENCE J. WOLF

R. P. WOOL

TING KAI WU

Applications of Polymer Spectroscopy

Edited by
EDWARD G. BRAME, JR.

ELASTOMER CHEMICALS DEPARTMENT
E. I. DU PONT DE NEMOURS & COMPANY
WILMINGTON, DELAWARE

ACADEMIC PRESS   New York   San Francisco   London   1978
A Subsidiary of Harcourt Brace Jovanovich, Publishers

PHYSICS

6060-5534√

ACADEMIC PRESS, INC.
111 Fifth Avenue, New York, New York 10003

United Kingdom Edition published by
ACADEMIC PRESS, INC. (LONDON) LTD.
24/28 Oval Road, London NW1 7DX

Library of Congress Cataloging in Publication Data

Main entry under title:

Applications of polymer spectroscopy.

Includes bibliographical references.
1. Polymers and polymerization--Spectra.
I. Brame, Edward Grant, Date
QC463.P5A66   547'.84   77-25728
ISBN 0-12-125450-X

PRINTED IN THE UNITED STATES OF AMERICA

Contents

4 Instrumental Characterization of Ethylene–Ethyl Acrylate–Carbon Monoxide Terpolymers

J. E. MCGRATH, L. M. ROBESON, T. E. NOWLIN, B. L. JOESTEN, AND E. W. WISE

5 Carbon-13 Spin Relaxation Parameters of Bulk Synthetic Polymers

RICHARD A. KOMOROSKI AND LEO MANDELKERN

6 NMR and Infrared Study of Thermal Oxidation of *cis*-1,4-Polybutadiene

ROBERT V. GEMMER AND MORTON A. GOLUB

7 NMR and Infrared Study of Photosensitized Oxidation of Polyisoprene

MORTON A. GOLUB, MARK L. ROSENBERG, AND ROBERT V. GEMMER

13 Characterization of Polymer Deformation and Fracture

DAVID K. ROYLANCE

14 Stress Mass Spectrometry of Nylon 66

MICHAEL A. GRAYSON AND CLARENCE J. WOLF

15 Characterization of Polymer Decomposition Products by Electron Impact and Chemical Ionization Mass Spectrometry

D. A. CHATFIELD, F. D. HILEMAN, K. J. VOORHEES, I. N. EINHORN, AND J. H. FUTRELL

16 Mass Spectrometry of Thermally Treated Polymers

G. J. MOL, R. J. GRITTER, AND G. E. ADAMS

List of Contributors

Numbers in parentheses indicate the pages on which the authors' contributions begin.

G. E. ADAMS (257), General Products Division, IBM Corporation, San Jose, California 95193

F. J. BOERIO (171), Department of Materials Science and Metallurgical Engineering, University of Cincinnati, Cincinnati, Ohio 45221

BERNARD J. BULKIN (121), Department of Chemistry, Polytechnic Institute of New York, Brooklyn, New York 11201

D. A. CHATFIELD (241), Flammability Research Center, Department of Materials Science and Engineering, University of Utah, Salt Lake City, Utah 84108

HUNG YU CHEN (7), Research Division, The Goodyear Tire and Rubber Company, Akron, Ohio 44316

M. M. COLEMAN (135), Polymer Science Section, Department of Materials Science and Engineering, Pennsylvania State University, University Park, Pennsylvania 16802

I. N. EINHORN (241), Flammability Research Center, Department of Materials Science and Engineering, University of Utah, Salt Lake City, Utah 84108

J. H. FUTRELL (241), Flammability Research Center, Department of Materials Science and Engineering, University of Utah, Salt Lake City, Utah 84108

ROBERT V. GEMMER* (79, 87), Ames Research Center, National Aeronautics and Space Administration, Moffett Field, California 94035

MORTON A. GOLUB (79, 87, 101), Ames Research Center, National Aeronautics and Space Administration, Moffett Field, California 94035

MICHAEL A. GRAYSON (221), McDonnell Douglas Research Laboratories, McDonnell Douglas Corporation, St. Louis, Missouri 63166

* Present address: American Cyanamid Company, Stamford, Connecticut 06904.

R. J. GRITTER (257), IBM Corporation, San Jose, California 95193

F. D. HILEMAN (241), Flammability Research Center, Department of Materials Science and Engineering, University of Utah, Salt Lake City, Utah 84108

HARVEY H. HOEHN (19), Central Research and Development Department, E. I. du Pont de Nemours & Company, Wilmington, Delaware 19893

B. L. JOESTEN (41), Research and Development Department, Union Carbide Corporation, Bound Brook, New Jersey 08805

RICHARD A. KOMOROSKI (57), Diamond Shamrock Corporation, Painesville, Ohio 44077

J. E. McGRATH (41), Chemistry Department, Virginia Polytechnic Institute and State University, Blacksburg, Virginia 24061

LEO MANDELKERN (57), Department of Chemistry and Institute of Molecular Biophysics, Florida State University, Tallahassee, Florida 32306

G. DAVID MENDENHALL (101), Battelle Laboratories, Columbus, Ohio 43201

G. J. MOL (257), IBM Corporation, San Jose, California 95193

RICHARD A. NATHAN (101), Battelle Laboratories, Columbus, Ohio 43201

T. E. NOWLIN (41), Research and Development Department, Union Carbide Corporation, Bound Brook, New Jersey 08805

DERICK W. OVENALL (19), Central Research and Development Department, E. I. du Pont de Nemours & Company, Wilmington, Delaware 19898

P. C. PAINTER (135), Polymer Science Section, Department of Materials Science, Pennsylvania State University, University Park, Pennsylvania 16802

L. M. ROBESON (41), Research and Development Department, Union Carbide Corporation, Bound Brook, New Jersey 08805

MARK L. ROSENBERG (87), Ames Research Center, National Aeronautics and Space Administration, Moffett Field, California 94035

DAVID K. ROYLANCE (207), Department of Materials Science and Engineering, Massachusetts Institute of Technology, Cambridge, Massachusetts 02139

W. O. STATTON* (185), Materials Science and Engineering, University of Utah, Salt Lake City, Utah 84112

ROGER S. STRINGHAM (1), Science Applications, Inc., Sunnyvale, California 94086

* Present address: 157 Mano Drive, Kula, Hawaii 96790.

MADELINE S. TOY (1), Science Applications, Inc., Sunnyvale, California 94086

K. J. VOORHEES (241), Flammability Research Center, Department of Materials Science and Engineering, University of Utah, Salt Lake City, Utah 84108

E. W. WISE (41), Research and Development Department, Union Carbide Corporation, South Charleston, West Virginia 25303

CLARENCE J. WOLF (221), McDonnell Douglas Research Laboratories, McDonnell Douglas Corporation, St. Louis, Missouri 63166

R. P. WOOL (185), Department of Metallurgy and Mining Engineering, University of Illinois, Urbana, Illinois 61801

TING KAI WU (19), Plastic Products and Resins Department, E. I. du Pont de Nemours & Company, Wilmington, Delaware 19898

Preface

Spectroscopy has become a powerful tool for the determination of polymer structure. With the current general use of computers and the newer methods being developed to handle polymers in examinations, an even more powerful approach is emerging for determinations. In response to the new developments a symposium was organized at the national American Chemical Society meeting in San Francisco to discuss them. Using the symposium as a basis to show the variety of spectroscopy methods available for the determination of polymer structure, this book was conceived to convey the ideas and concepts presented and discussed there. However, it is not a proceeding of the symposium. It contains new and updated material in addition to that presented at the meeting and in its organization reflects the various applications of polymer spectroscopy.

The book is divided into three general areas of spectroscopy. The first area is devoted to the applications of nuclear magnetic resonance (NMR) spectroscopy. Not only is there the expected coverage of the use of carbon-13 NMR but also there is coverage of the use of proton NMR and even the lesser used fluorine-19 NMR. The second general area is infrared spectroscopy. The coverage here involves the use of the new and powerful Fourier transform method as well as the dynamic method of handling the examination of polymer films. The third general area is devoted to applications of mass spectroscopy. The discussion covers a broad range of topics from the usual characterization of decomposition products both by direct and indirect means to the lesser known characterization of these products by stressing the polymer.

In addition to the three general areas cited above there is some discussion on the use of chemiluminescence, Raman, and electron spin resonance (ESR) methods. The last covers the application of the ESR method to the examination of polymers under stress. Hence, with the inclusion of the ESR method, along with the two chapters devoted to the use of infrared and mass spectroscopy methods in examining polymers under stress, this book offers a most complete account on the subject of polymer spectroscopy.

Finally, I wish to thank all of the contributors to this book for their support and for their willingness to provide their latest findings to make the book as useful and as complete as possible. Thanks is also given to Dr. Golub who originally suggested publishing the material presented at the symposium on polymer spectroscopy at the San Francisco meeting. This book is a consequence of his suggestion.

1

Fluorine-19 and Carbon-13 NMR Spectra of Polyperfluorobutadiene

MADELINE S. TOY and ROGER S. STRINGHAM

SCIENCE APPLICATIONS, INC.
SUNNYVALE, CALIFORNIA

I. INTRODUCTION

The application of high-resolution nuclear magnetic resonance (NMR) spectroscopy to determine polymer structures has been demonstrated and reviewed as a useful analytic tool over a decade ago [1,2]. Considerable data on characterizing microstructures of a variety of polymers are now available by ^{1}H NMR and ^{13}C NMR spectra and by ^{19}F NMR for fluoropolymers [1–5].

This chapter describes the ^{19}F NMR and ^{13}C NMR spectra of polyperfluorobutadiene melt and solution below 200°C. The microstructure of polyperfluorobutadiene as deduced by NMR spectra and the comparison to infrared (ir) analysis [6,7] are discussed.

II. EXPERIMENTAL TECHNIQUE

The polyperfluorobutadienes were prepared in bulk at ambient temperature using 1.0 mole percent perfluorodialkyl peroxide as initiator under photolysis. The molecular weight of the solid polymer is about 10,000 with

1

T_m 100–130°C. The method of preparation and characterization were previously reported [8,9]. Perfluorobutadiene monomer and bis(trifluoromethyl) peroxide were obtained from PCR and bis(perfluoro-t-butyl) peroxide [10] was prepared in our laboratory [11,12].

All NMR polymer samples were prepared as neat or as 10% solutions in molten camphor, and used 12 mm outside diameter, Pyrex NMR tubes. Spectra were obtained from Varian XL-100 NMR spectrometer operating at 94.1 MHz for ^{19}F NMR and 25.2 MHz for ^{13}C NMR. Hexafluorobenzene was used as an external standard for ^{19}F NMR spectra. The chemical shifts were converted with respect to fluorotrichloromethane. The deuterated glycerol was used as an external lock. For ^{13}C NMR spectra trifluoromethyl iodide was used as an external standard and the chemical shifts were converted to tetramethylsilane. The deuterated dimethyl sulfoxide was used as an external lock. The probe of the Varian XL-100 spectrometer was modified by pulsed energy input for ^{13}C NMR with ^{19}F decoupling and was irradiated with a broad frequency (12,000 Hz band) to decouple ^{19}F nuclei from ^{13}C nuclei.

III. RESULTS AND DISCUSSION

Figure 1 shows the ^{19}F NMR spectrum of a solution of 0.1 gm polyperfluorobutadiene in 0.9 gm of molten camphor at 160°C. It should be pointed out that the same polymer (neat) was also observed from 100–190°C. Although the resolution of the spectra improves with temperature, all spectra show the same two major peaks of relative areas in 2:1 ratio as Fig. 1. The chemical shifts are at −113.5 and −150.0 ppm in 2:1 ratio and are assigned as —CF$_2$— (not terminal —CF=CF$_2$) and —CF= (not tertiary

—CF), respectively [13,14]. Thus the microstructure of polyperfluorobutadiene consists mainly of 1,4-moieties.

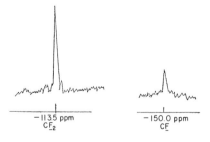

FIG. 1 ^{19}F NMR of a solution of 10% polyperfluorobutadiene in molten camphor. (Note that compounds distinguished in the figures by an underline are distinguished in the text by boldface.)

TABLE 1

^{19}F NMR of Polyperfluorobutadiene Melt[a] at 160°C

Polymer	Chemical shift (ppm from $CFCl_3$)	Relative area	Assignment	No. of F	Relative area — No. F ($\times 100$)	No. of molecules[b]
$\overset{a}{(CF_3)_3}CO$	−65.5	0.11	a	9	1	4 — $(CF_3)_3COOC(CF_3)_3$
$\overset{b}{(CF_3)_3}CO$	−67.7	0.64	b	9	7	
$\overset{c}{C}F_2$	−112.0	0.156	c	2	7	
$\overset{d}{C}F_2$	−115.5	1.85	d	2	92	
$\overset{e}{C}F$	−132.4	0.075	e	1	7	51 — $CF_2CF{=}CFCF_2$
$\overset{f}{C}F$	−151.7	0.925	f	1	92	
$\overset{g}{C}F$	−153.1	0.075	g	1	7	

[a] 1% $(CF_3)_3COOC(CF_3)_3$ as initiator.

[b] Approximate number of molecules for $(CF_3)_3COOC(CF_3)_3 = \frac{1}{2}(1 + 7) = 4$. Approximate number of molecules for $CF_2CF{=}CFCF_2 = \frac{1}{4}(7 + 92 + 7 + 92 + 7) = 51$.

The bulk polymer initiated by bis(trifluoromethyl) peroxide catalyst previously analyzed by ir spectroscopy showed a sharp decrease at 5.6 μm for the band due $-CF{=}CF_2$ groups [15] (indicating 1,2-mioties of polyperfluorobutadiene) with decreasing initiator concentration to 1% [16]. This discussion shows the microstructure as mainly 1,4-moieties and also illustrates the fact that NMR spectroscopy is a more powerful tool for microstructural elucidation of polyperfluoropolyene than ir due to the ir inactive trans-1,4-moieties [17]. In other words, the ir absorption for $-CF{=}CF-$ groups (indicating 1,4-moieties) at 5.8 μm is limited to cis-1,4-moieties. Thus when the trans-1,4-units in polyperfluorobutadiene outnumber the cis-1,4-, the ratio of 1,2- and 1,4-units based on ir analysis alone can be misleading.

Table I summarizes ^{19}F NMR data of 1% bis(perfluoro-t-butyl) peroxide initiated polyperfluorobutadiene melt at 160°C. The two major peaks are at -115.5 ppm assigned for CF_2^d and -151.7 ppm assigned for CF^f in 2:1 ratio. One concludes that the microstructure consists mainly of 1,4-moieties but with additional peaks at high field of the -65 to -68 ppm region for perfluoro-t-butoxy groups from the initiator. The average tentative structure deduced from Table I is suggested as follows:

$$\begin{array}{c} \mathrm{CF{=}CF_2} \qquad\qquad\qquad \mathrm{OC(CF_3)_3} \\ \quad| \qquad\qquad\qquad\qquad\qquad | \\ \mathrm{(\overset{a}{CF}_3)_3CO\overset{}{C}FCF_2(CF_2C\overset{d}{F}{=}\overset{f}{C}FCF_2\!\!\underset{\overline{n}}{)}\,[CF_2\overset{e}{C}F\overset{g}{C}FCF_2(CF_2CF{=}CFCF_2)_n]_3\,CF_2CF{=}CFCF_2OC(CF_3)_3} \\ | \\ \mathrm{(CF_2CF{=}CFCF_2)_nCF_2CF{=}CFC\overset{c}{F}_2OC(\overset{b}{C}F_3)_3} \end{array}$$

where $\sum n \simeq 42$.

The ^{13}C NMR spectrum of polyperfluorobutadiene melt (Fig. 2b) shows a triplet peak at 149.5 ppm and a doublet peak at 108 ppm from tetramethylsilane as assigned to $\mathbf{CF_2}$ and $\mathbf{CF}$ respectively. Further splitting of the doublet peak indicates long range coupling with $\mathbf{CF_2}$.

The probe modification for ^{13}C NMR with ^{19}F decoupling is described in Section II. Figure 2a is a ^{13}C NMR spectrum with ^{19}F decoupling. The equal amount of the two types of carbon at 149.5 ppm and 108 ppm are assigned as $\mathbf{CF_2}$ and $\mathbf{CF}$. The selective decoupling as shown by Fig. 2a confirms the number of fluorine atoms attached to the particular carbon of the undecoupled (Fig. 2b) and also establishes the identical relationship between the resonance frequencies of the two types of carbon as 810 ± 10 Hz between $\mathbf{CF_2}$ and $\mathbf{CF}$.

In conclusion, the microstructure of polyperfluorobutadiene from ^{13}C NMR spectra complements and confirms the ^{19}F NMR as consisting mainly of 1,4-moieties. Based on the low intensity of the ir absorption at 5.8 μm, the ir inactive trans-1,4-moiety is suggested as the predominant configuration.

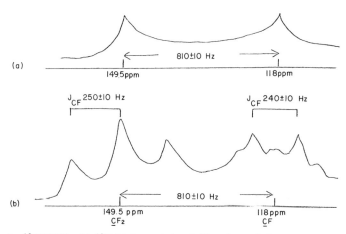

FIG. 2 ^{13}C NMR with ^{19}F (a) decoupled and (b) undecoupled spectra of polyperfluoro-butadiene melt.

ACKNOWLEDGMENTS

We wish to acknowledge the Air Force Office of Scientific Research for support of this work under Contract F44620-76-C-0027 and the assistance of Mr. L. Cary for NMR spectra and the modification of the probe of the Varian XL-100 spectrometer.

REFERENCES

1. F. A. Bovey and G. V. D. Tiers, *Fortschr. Hochpolym. -Forsch.* **3**, 139 (1963).
2. K. C. Ramey and W. S. Brey, *J. Macromol. Sci., Rev. Macromol. Chem.* **1** (2), 263 (1967).
3. G. C. Levy and G. L. Nelson, "Carbon-13 Nuclear Magnetic Resonance for Organic Chemists." Wiley (Interscience), New York, 1972.
4. J. B. Stothers, "Carbon-13 NMR Spectroscopy." Academic Press, New York, 1971.
5. L. A. Wall (ed.) "Fluoropolymers." Wiley (Interscience), New York, 1971.
6. M. S. Toy and D. D. Lawson, *J. Polym. Sci. Part B* **6**, 639 (1968).
7. M. S. Toy and J. C. DiBari, *Polymer* **14**, 327 (1973).
8. M. S. Toy and J. M. Newman, *J. Polym. Sci., Part A* **17**, 2333 (1969).
9. M. S. Toy, "Photochemistry of Macromolecules" (R. F. Reinisch, ed.), pp. 135–144. Plenum, New York, 1970.
10. D. E. Gould, C. T. Ratcliff, L. R. Anderson, and W. B. Fox, *Chem. Commun.*, 216 (1970).
11. M. S. Toy and R. S. Stringham, *J. Fluorine Chem.* **5**, 481 (1975).
12. M. S. Toy and R. S. Stringham, *J. Fluorine Chem.* **7**, 375 (1976).
13. J. W. Emsley, J. Feeney, and L. H. Sutcliffe, "High Resolution Nuclear Magnetic Resonance Spectro," Vol. II. Pergamon, Oxford, 1966.
14. C. H. Dungan and J. R. VanWazer, "Compilation of Reported ^{19}F NMR Chemical Shifts." Wiley (Interscience), New York, 1970.
15. P. Tarrant, *Polym. Prepr. Amer. Chem. Soc., Div. Polym. Chem.* **12** (1), 391 (1971).
16. M. S. Toy, *Polym. Prepr., Amer. Chem. Soc., Div. Polym Chem.* **12** (1), 385 (1971).
17. J. K. Brown and K. J. Morgan, *Adv. Fluorine Chem.* **4**, 264 (1965).

2

Carbon-13 NMR Spectra of Polyalkenamers

HUNG YU CHEN

REASEARCH DIVISION
THE GOODYEAR TIRE AND RUBBER COMPANY
AKRON, OHIO

I. INTRODUCTION

In characterization of hydrocarbon polymers, both proton and carbon-13 nuclear magnetic resonance (NMR) techniques are very effective. However, carbon-13 NMR has several advantages over proton NMR in the study of polymer chain structures. One is that the carbon-13 chemical shift is more sensitive to the effect from neighboring groups. Carbon chemical shifts of organic compounds cover a range of approximately 600 ppm as compared with under 20 ppm for the proton chemical shifts. As a result, carbon-13 NMR is more responsive to small changes in polymer chain structures and is, therefore, frequently used for studying chain sequence distribution of polymers. The other advantage is the regularity of the carbon-13 chemical shift relative to the chemical structure of carbon nuclei. In most instances, one can predict the carbon-13 chemical shift from the chemical structure of polymers with reasonable accuracy. The carbon-13 chemical shift of a

7

given carbon in a family of closely related polymers can be correlated by an empirical equation with a set of parameters, each corresponding to the individual contribution of a particular structural unit in the polymer [1,2]. This linear correlation can permit peak assignments in the sequence distribution analysis of polymers or copolymers having the chemical structures closely resembling the model polymers.

The polymer system used in our work is polyalkenamers, which represent a family of polymers with the following general formula:

$$\{CH_2-(CH_2)_n-CH=CH\}_x$$

The polymers are prepared from the olefin metathesis [3a–c] of cycloolefins:

$$\binom{CH=CH}{CH_2-(CH_2)_n} \xrightarrow[\text{Catalyst}]{\text{W or Mo}} \{CH_2-(CH_2)_n-CH=CH\}_x$$

With $n = 1, 2, 3, \ldots$, etc., the corresponding names of the polymers are polybutenamer, polypentenamer, polyhexenamer, . . . , etc.

Each of the polymers has a linear structure with the vinylene double bonds regularly spaced along the polymer chain. Each pair of adjacent double bonds is separated by a group of methylene units. Since the vinylene double bonds can be either in cis or trans configuration, the polymer can have the cis and trans double bonds arranged in many different ways along the polymer chain. Thus, a given methylene unit can be in any one of the following possible environments:

$$-(cis)-(CH_2)_n-(cis)-$$
$$-(cis)-(CH_2)_n-(trans)-$$
$$-(trans)-(CH_2)_n-(trans)-$$

The polymers are therefore ideal model compounds for the various methylene structures shown above. We can use these polymers to determine the effects of *cis*- and *trans*-vinylene double bonds relative to the methylene carbon chemical shifts of above structures. Subsequently, we can obtain the additivity parameters of the methylene carbon chemical shifts due to the vinylene double bonds. The parameters thus obtained in turn can be used for making peak assignments in diene polymers or copolymers.

II. EXPERIMENTAL TECHNIQUE

The proton decoupled carbon-13 NMR spectra of the polymers were obtained at ambient temperature with a Varian CFT-20 NMR spectrometer or a Varian XL-100 NMR spectrometer equipped with FT accessories.

Deutero-chloroform was used as the solvent for the polymer solutions. Tetramethylsilane (TMS) was added to the solutions for the internal standard. All chemical shifts of the spectral peaks were measured in ppm relative to TMS. Concentrations of the polymer solutions were about 3–10%. 10-mm and 15-mm spinning tubes were used in the CFT-20 and XL-100 spectrometers, respectively.

Both spectrometers are equipped with 16 K computers which provide a 1 Hz per data point resolution at 4000 Hz spectral width for the spectra. The significance of the last digit of the chemical shift value in ppm is limited to this 1-Hz resolution for all spectral data used in this article. The other FT parameters used for the spectrometers were optimized for qualitative spectra. For resolving the methylene peaks in polybutenamer we used the computer resolution of 0.2 Hz per data point. The overall resolution of the instrument measurement was limited by the magnet homogeneity of the CFT-20, which is about 0.3 Hz.

III. RESULTS

The polyalkenamers included in this study are those of $n = 1$, $n = 2$, $n = 4$, and $n = 5$; they are, respectively, polybutenamer, polypentenamer, polyheptenamer, and polyoctenamer. For the purpose of this study, only the carbon-13 chemical shifts of the methylene groups are included. In the assignments of the methylene peaks, polypentenamer is used as a representative for the discussion in detail [4].

A. Polypentenamer

The carbon-13 NMR spectrum of the methylene units in polypentanamer with cis/trans ratio being 40/60, is given in Fig. 1. In polypentenamer, there are three methylene units between the two adjacent vinylene double bonds.

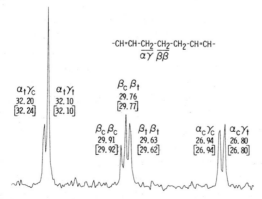

FIG. 1   Methylene region of the carbon-13 NMR spectrum of polypentenamer. $CDCl_3$ solution is at ambient temperature. Chemical shifts are in ppm units from TMS. Calculated values are in brackets.

If one considers only the effect due to the first neighboring double bonds, there will be three possible chain environments in which the methylene units may be situated, namely:

$$—(cis)—CH_2CH_2CH_2—(cis)— \qquad \text{(I)}$$
$$—(cis)—CH_2CH_2CH_2—(trans)— \qquad \text{(II)}$$
$$—(trans)—CH_2CH_2CH_2—(trans)— \qquad \text{(III)}$$

The two methylene units α position to the *cis*-vinylene double bonds in (I) would have the same carbon-13 chemical shift. Therefore, two peaks are expected from the methylene units in (I). Similarly, there should be two peaks from the methylene units in (III). Three peaks are expected from (II) because the methylene units in (II) are all different. Thus, from (I), (II), and (III) there should be seven peaks possible in the methylene region of the carbon-13 NMR spectrum for the polymer. All seven peaks were observed experimentally from the spectrum of the polymer shown in Fig. 1. The assignments of the peaks were made by comparing the change of the peak intensities with respect to the cis/trans ratios of three polypentenamer samples, one with a low trans content, the second with a high trans content, and the third with an approximately equal amount of *cis*- and *trans*-vinylene double bonds in the polymer. The reasoning behind this is that, for a poly-pentenamer with a high trans content, the trans double bonds are expected to be mainly in the structural unit (III). Therefore, the major carbon-13 spectral lines in the methylene region of the spectrum for the polymer will consist of those contributed from that structural unit. On the other hand, the situation is just the reverse for a polypentenamer with a low trans content (or a high cis content), and (I) will be the major contributor. For polypentenamer having about equal cis and trans double bonds, all three structural units (I), (II), and (III) contribute. Based on this technique, all seven methylene peaks were assigned as given in Fig. 1. In the figure, each peak and the methylene unit attributes to the peak are labeled by a double Greek letter designation, with subscripts of either c or t for cis or trans double bond, to indicate how many carbon–carbon bonds the cis and trans double bonds are away from the methylene unit in question. The measured chemical shifts in ppm unit are also given for each of the peaks with the calculated values in bracket in the figure. The calculated chemical shifts are from an additivity formula which will be discussed later.

The spectrum consists of three absorption regions. The closely spaced peaks at the low field side (large ppm value) are from the methylene units α position to the *trans*-vinylene double bond. The additional effect of the cis or trans double bond in the γ position introduces the small difference in the chemical shift so that two closely spaced peaks were observed. The trans double bond in the γ position gives an upfield shift as compared to a downfield shift for the trans double bond in the α position. The cis double bond in the α position causes an upfield shift of the methylene chemical shift. With the additional effect of the double bonds in the γ position, two closely spaced peaks were also observed at the high field side of the spectrum for the methylene units next to a cis double bond. Here again, the trans double bond in the γ position gives an upfield shift as in the case of the methylene unit next to a trans double bond. The three peaks in the middle region of the spectrum are from the methylene units that are in the β position with respect to the double bonds on either sides. The trans double bond in the β position causes an upfield shift while the cis double bond causes a downfield shift. Qualitatively, we found the following directions of shift due to cis and trans double bonds in different positions:

$$\text{Double bonds in } \alpha \text{ position} \begin{cases} \text{trans} \longleftarrow \\ \text{cis} \longrightarrow \end{cases}$$

$$\text{Double bonds in } \beta \text{ position} \begin{cases} \text{trans} \longrightarrow \\ \text{cis} \longleftarrow \end{cases}$$

A quantitative treatment of the shifts due to the double bonds in different positions will be given in the latter part of the article.

B. Polyheptenamer

In Fig. 2, the carbon NMR spectrum of the methylene units of polyheptenamer is given. The assignments of the peaks were made as described in the polypentenamer case. Four groups of peaks were observed, which

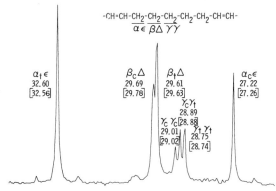

FIG. 2   Methylene region of the carbon-13 NMR spectrum of polyheptenamer. CDCl$_3$ solution is at ambient temperature. Chemical shifts are in ppm units from TMS. Calculated values are in brackets.

are assigned to the three types of methylene units, labeled $\alpha\varepsilon$, $\beta\Delta$, and $\gamma\gamma$, according to the convention mentioned previously. The first type of methylene units, $\alpha\varepsilon$, has cis and trans double bonds in the α position. The relatively large upfield and downfield shifts due to cis and trans double bonds cause the methylene peaks to be at high and low field sides (27.22 ppm and 32.60 ppm), respectively. Since both peaks at 27.22 ppm and 32.60 ppm are singlets, it indicates that the effects due to trans and cis double bonds in the ε position to the methylene chemical shift must be the same. In other words, the methylene unit can no longer sense the difference between cis and trans double bonds five carbon–carbon bonds away. Because only two peaks were observed for the $\beta\Delta$ methylene units at 29.61 ppm and 29.69 ppm and β_t and β_c are known to be different as already observed in the polypentenamer case, the same is apparently true for the Δ position, which is four carbon–carbon bonds away. Three peaks were observed at 28.75 ppm, 28.89 ppm, and 29.01 ppm for $\gamma\gamma$ methylene units. These peaks correspond to the three possible combinations, $\gamma_c\gamma_c$, $\gamma_c\gamma_t$, and $\gamma_t\gamma_t$. This indicates that the cis and trans effects are distinguishable in the γ position, i.e., they are distinguishable up to three carbon–carbon bonds away.

C. Polyoctenamer

The spectrum of methylene units in polyoctenamer containing both cis and trans double bonds consists of four groups of peaks as shown in Fig. 3. The assignments of these peaks were made as described before. The general

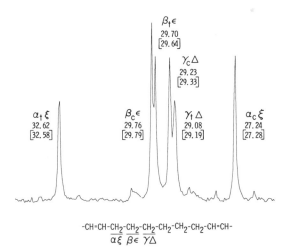

$$-CH=CH-CH_2-CH_2-CH_2-CH_2-CH_2-CH_2-CH=CH-$$
$$\underline{\alpha\xi}\ \ \underline{\beta\epsilon}\ \ \overline{\gamma\Delta}$$

FIG. 3   Methylene region of the carbon-13 NMR spectrum of polyoctenamer. CDCl$_3$ solution is at ambient temperature. Chemical shifts are in ppm units from TMS. Calculated values are in brackets.

features of the spectrum are consistent with those just described in the preceding sections. That is, the different effects of cis and trans double bonds can only be distinguished up to the third carbon–carbon bond from them. As a result, two singlets corresponding to $\alpha_t\xi$ and $\alpha_c\xi$ were observed at 32.62 ppm and 27.24 ppm, respectively. Two closely spaced peaks at 29.76 ppm and 29.70 ppm for $\beta_c\varepsilon$ and $\beta_t\varepsilon$ as well as two closely spaced peaks at 29.23 ppm and 29.08 ppm for $\gamma_c\Delta$ and $\gamma_t\Delta$ were observed in the middle portion of the spectrum.

D. Polybutenamer

Polybutenamer is the only member of the polyalkenamers that gives an anomalous spectrum. The methylene units in the polymer exhibit only two peaks as shown in Fig. 4. In the polybutenamer structure, there are two methylene units between two adjacent double bonds. Consequently, there should be observed four peaks corresponding to $\alpha_t\beta_t$, $\alpha_t\beta_c$, $\alpha_c\beta_t$, and $\alpha_c\beta_c$. The fact that only two peaks were observed experimentally requires either $\alpha_t = \alpha_c$ or $\beta_t = \beta_c$. Judging from the separation of the peaks, 5.31 ppm, it appears more likely that the discrepancy is due to the condition of $\beta_t = \beta_c$. From the other members of the polyalkenamers studied in this article, the average difference between β_t and β_c was found to be 0.07 ppm. This separation could easily be resolved under the experimental conditions used for the spectrum. Attempts to resolve the peaks under high resolution conditions as well as using the aromatic solvent effect were not successful. More discussion on this point will be given later. This anomalous phenomenon has an important bearing in practical applications. Polybutenamer has the chemical structure of 1,4-polybutadiene. When carbon-13 NMR spectroscopy is used for studying the chain sequence distribution of *cis*- and

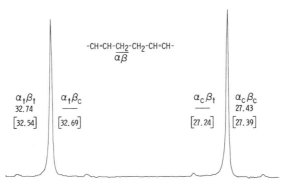

FIG. 4   Methylene region of the carbon-13 NMR spectrum of polybutenamer. $CDCl_3$ solution is at ambient temperature. Chemical shifts are in ppm units from TMS. Calculated values are in brackets.

trans-butadiene units in the polymer, the presence of only two peaks foɪ
the methylene units could lead one to the erroneous conclusion that the
polybutadiene contains only cis and trans blocks. That may occur even
though it is actually a random copolymer of *cis*- and *trans*-1,4-polybutadiene
[5a–d].

IV. DISCUSSION

As mentioned earlier, the chemical shifts of carbon-13 nuclei in different
chemical environments can usually be correlated in an additive fashion
by empirical equations. The carbon-13 chemical shifts for the methylene
units in the polyalkenamers may be written as [6]

$$\delta \text{ppm from TMS} = B + \sum A_f$$

where B is a constant and A_f are additivity parameters for the effects of
the double bonds at different positions with respect to the methylene unit
under consideration. Using the assumption as mentioned in the polypentenamer case that only the first double bonds on both sides are to be
considered significant, there are two A_f parameters required for each type
of methylene unit in the equation. The notation for the parameters, A_f, is
the same one used for the designation of the spectral peaks with slight
modification. A_f is denoted by a Greek letter with subscript c or t for cis or
trans double bond. The Greek letter itself indicates how many carbon–carbon
bonds the methylene unit is from the double bond, i.e., $\alpha, \beta, \gamma \ldots$, etc., mean
one, two, three, $\ldots$, etc., carbon–carbon bonds. The physical meaning
of the equation may be visualized as follows. The polyalkenamers are
considered to be derived from a basic polymer, polymethylene. The B constant in the equation is the carbon-13 chemical shift of the methylene units
in the basic polymer, polymethylene. Each member of the polyalkenamers
is equivalent to the basic polymer with addition of equally spaced double
bonds. The change of the methylene carbon-13 chemical shift because of
the addition of the double bonds is corrected by the two A_f parameters,
one for each side of the first neighboring double bond. The convention is
that when a double bond causes the methylene chemical shift to shift downfield, the additivity parameter A_f for the effect is positive, and vice versa.
In this way, each type of methylene unit in polyalkenamers can be labeled
by a double Greek letter designation, which also denotes the two additivity
parameters in the additivity equation. The values of individual additivity
parameters are extracted from the spectra. A given additivity parameter
usually appears in different members of the polyalkenamers. The value of
each additivity parameter is chosen so that it fits best for all of the poly-

alkenamer spectra involved. The values of the additivity parameters A_f thus obtained are listed in Table I. From these additivity parameters, the chemical shifts of the methylene units in the polyalkenamers reported here were calculated using the above equation. The calculated values are given in brackets alongside each measured value for the methylene peaks in the different figures. The excellent agreement between the measured and calculated values for the methylene peaks confirms the regularity of the additivity nature of the carbon-13 chemical shift of the methylene units in relation to their chemical environments. The only member of the polyalkenamers which does not fit this pattern is polybutenamer. As mentioned previously, polybutenamer has two methylene units between two adjacent double bonds. It is expected that both α and β effects are observed as in polypentenamer. The problem that only two peaks are observed may be attributed to the assumption made previously that only the first neighboring double bonds are included in the additivity effects. This assumption is apparently valid, when there are more than two methylene units between the two adjacent double bonds. But when there are less methylene units between the double bonds, the second-order effect from the double bonds other than the first neighboring double bonds becomes sufficiently significant so that it can no longer be ignored. In the polybutenamer case, this second-order effect combines with the β effect to cause the net β effect between β_c and β_t to be identical [4]. The explanation appears reasonable. However, no experimental data are available at this time to substantiate it.

In summary, we have correlated the methylene carbon-13 chemical shifts in polyalkenamers with respect to their chemical environments, i.e.,

TABLE I

Additivity Equation and Additivity Parameters for the Methylene Carbon-13 Chemical Shifts of Polyalkenamers[a]

Additivity formula	$\delta_{TMS} = B + \sum\limits_f A_f$
Additivity parameters	$B = 29.76$
	$\alpha_c = -2.45 \qquad \alpha_t = 2.85$
	$\beta_c = 0.08 \qquad \beta_t = -0.07$
	$\gamma_c = -0.37 \qquad \gamma_t = -0.51$
	$\Delta = -0.06$
	$\varepsilon = -0.05$
	$\xi = -0.03$

[a] Units in ppm from TMS. $CDCl_3$ solution at ambient temperature.

their relative positions with respect to the first neighboring double bonds by a set of additivity parameters given in Table I. These parameters correlate the methylene chemical shifts of polyalkenamers well when $n = 2$, 4, and 5. It is believed that the parameters will work well for $n > 5$ as well.

In addition to demonstrating the regularity of the additivity property of the carbon-13 chemical shifts of polyalkenamers, the above additivity parameters may be used in the elucidation of polymer chain sequence distributions. For instance, polyalkenamers may be considered as copolymers of 1,4-butadiene such as polyoctenamer, which has the repeating structure:

$$\text{-}[\text{CH}_2\text{--CH}_2\text{--CH}_2\text{--CH}_2\text{--CH}_2\text{--CH}\text{=CH--CH}_2\text{-}]_x$$

This is equivalent to an alternate copolymer of 1,4-butadiene and tetramethylene monomeric units.

In order to show the practical use of the additivity equation with the additivity parameters, we show a spectrum of partially hydrogenated 1,4-polybutadiene ($\sim 30\%$ hydrogenated) [7]. The methylene region of the carbon-13 spectrum of the polymer is given in Fig. 5. The spectrum resembles that of polyoctenamer. That is because a triad block with an isolated hydrogenated butadiene unit in the middle is the basic structure of polyoctenamer. The whole spectrum is the result of superimposed spectra of 1,4-polybutadiene, polyoctenamer, and hydrogenated butadiene blocks. With the peak assignments made, it is a routine task to obtain the relative peak intensities for each of the contributing structures in order to provide the basic data for the chain sequence distribution calculations.

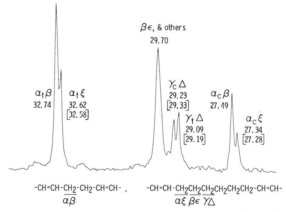

FIG. 5 Methylene region of the carbon-13 NMR spectrum of partially hydrogenated 1,4-polybutadiene. CDCl$_3$ solution is at ambient temperature. Chemical shifts are in ppm units from TMS. Calculated values are in brackets.

REFERENCES

1. J. B. Stothers, "Carbon-13 NMR Spectroscopy," Chapter 3. Academic Press, New York, 1972.
2. G. C. Levy and G. L. Nelson, "Carbon-13 NMR for Organic Chemists," Chapter 3. Wiley (Interscience), New York, 1972.
3a. N. Calderon, H. Y. Chen, and K. W. Scott, *Tetrahedron Lett.* **34**, 3327 (1967).
3b. N. Calderon, E. A. Ofstead, J. P. Ward, W. A. Judy, and K. W. Scott, *J. Amer. Chem. Soc.* **90**, 4133 (1968).
3c. N. Calderon and H. Y. Chen, U. S. Patent 3,535,401.
4. H. Y. Chen, *J. Polym. Sci. Part B* **12**, 85 (1974).
5a. V. D. Mochel, *J Polym. Sci. Part A-1* **10**, 1009 (1972).

5b. J, Furukawa, E. Kobayashi, T. Kawagoe, N. Katsuki, and M. Imanari, *J Polym. Sci.* **11**, 245 (1973).
5c. A. D. H. Clague, J. A. M. van Brockhaven, and J. W. deHaan, *J Polym. Sci. Part B* **11**, 305 (1973).
5d. E. R. Santee, V. D. Mochel, and M. Morton, *J Polym. Sci. Part B* **11**, 452 (1973).
6. G. Gatti and A. Carbonaro, *Makromol. Chem.* **175**, 1627 (1974).
7. H. Y. Chen, *J. Polym. Sci. Part B* **15**, 271 (1977).

3

Proton and Carbon-13 NMR Spectra
of Ethylene–Carbon Monoxide Copolymers*

TING KAI WU

PLASTIC PRODUCTS AND RESINS DEPARTMENT
E. I. DU PONT DE NEMOURS & COMPANY
WILMINGTON, DELAWARE

DERICK W. OVENALL and HARVEY H. HOEHN

CENTRAL RESEARCH AND DEVELOPMENT DEPARTMENT
E. I. DU PONT DE NEMOURS & COMPANY
WILMINGTON, DELAWARE

I. INTRODUCTION

Copolymerization of ethylene (E) with carbon monoxide (CO) has been reported many years ago [2]. Infrared spectra revealed that E–CO

* See Wu [1].

copolymers are polyketones [2]. Experimental evidence to date shows that CO forms alternating sequences with E in the polymer chain such as —$E_m COE_n COE_p$— (m, n, and p are integers ≥ 1) [3,4]. Even though the distribution of E sequences can, in principle, be determined by analyzing the dibasic acids derived from the nitric acid oxidation products of E–CO polymers [2], the quantitative validity of this analysis is obviously dependent on the selectivity and completeness of the oxidation reaction. Previous parts of this series demonstrated the utility of high resolution nuclear magnetic resonance (NMR) spectroscopy for investigating the micro-structure of a variety of ethylene copolymers [5,6]. In this work, we applied proton and carbon-13 techniques as nondestructive methods for quantitative analysis of monomer sequence distribution in E–CO copolymerization. End groups and short chain branching were also observed. Moreover, the polyethylene segments of these E–CO polymers, like other ethylene co-polymers, exhibit conformational transitions in solution.

II. EXPERIMENTAL TECHNIQUE

E–CO copolymers were prepared by high-pressure free radical poly-merization of E and CO using a procedure similar to that reported previously [2,5]. The conversions were about 10%. Copolymer compositions were determined by elemental analyses of carbon, hydrogen and oxygen, and by NMR. Vapor phase and membrane osmometry and gel permeation chromatography were used to characterize the E–CO polymers. Their number-average molecular weights were found to exceed 11,800 in all cases. The proton NMR spectra were obtained on Varian A-60 and HR-220 NMR spectrometers. Carbon-13 spectra were recorded at 22.63 MHz by using a Bruker HFX-90 NMR spectrometer equipped with a Digilab NMR 3 Fourier transform accessory. Preparation of the sample solutions and spectral measurements were carried out in accordance with the previously described procedures [5–7]. The proton and carbon-13 chemical shift values were measured from the internal references, hexamethyl disiloxane (HMDS), and tetramethyl silane (TMS), respectively.

III. RESULTS AND DISCUSSION

In our previous studies of monomer sequence distributions in ethylene copolymerizations, spectral analysis was frequently complicated by addi-tional resonance lines arising from stereoregularity of the vinylic comono-mers. This complication is no longer present in the case of E–CO polymers because of the nonstereospecific nature of CO. Moreover, the fact that a CO unit must alternate with an E unit during copolymerization limits the number of different comonomer sequence placements. This further simplifies spectral assignments.

A. Proton Spectra

Figure 1 depicts the 60 MHz and 220 MHz proton spectra of an E–CO polymer dissolved in deuterated chloroform ($CDCl_3$) at 60°. The methylene proton resonances cover the region from 2.7 ppm to 1.2 ppm. In order to aid our spectral analysis we examined the spectra of aliphatic ketones and diones and those of the analogous ethylene–sulfur dioxide ($E–SO_2$) copolymers. Our assignments are summarized in Table I. It is seen that the shieldings of the methylene protons are dependent on their proximity to the neighboring carbonyl groups. They decrease in order of the positions δ, $\gamma > \beta > \alpha$ of the structure

$$-\overset{\alpha}{COCH_2}\overset{\beta}{CH_2}\overset{\gamma}{CH_2}\overset{\delta}{CH_2}-$$

In fact, the chemical shift of the methylene protons in the 1,4-dione structure $-COCH_2CH_2CO-$ is 2.6 ppm and this is, to a good approximation, the simple sum of α and γ effects. Even though additional lines are resolved in the 220 MHz spectrum, some of these are redundant and not useful for structural analysis. From structural considerations it is clear that the total intensity of the β methylene peak centered at about 1.5 ppm must be identical to that of the α methylene multiplet at about 2.3 ppm for copolymers of all

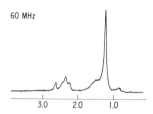

FIG. 1   60 MHz and 220 MHz proton spectra of an E–CO copolymer (21.8 mol % CO) in $CDCl_3$ at 60°

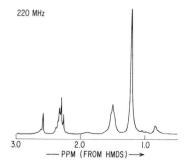

TABLE I

Proton Spectral Assignments for Ethylene–Carbon Monoxide Copolymers

Proton chemical shifts[a]		Protons[b]	Dyads, triads, etc.	Designation
0.809,	0.834	—CH₂CH₂CH₂**CH₃**		Alkyl methyl
0.959,	0.977, 0.990	—COCH₂**CH₃**		Keto methyl
1.21		CH₂CH₂**CH₂**CH₂CH₂	EE	γ methylene
1.50		COCH₂**CH₂**CH₂CH₂	COEE	β methylene
2.29		CH₂CH₂CH₂CH₂**COCH₂**CH₂CH₂CH₂CH₂	EECOEEE ⎫	
2.35		COCH₂CH₂CH₂**COCH₂**CH₂CH₂CH₂CH₂	COEOEEE ⎬	α methylene
		CH₂CH₂CH₂CH₂**COCH₂**CH₂CH₂CH₂CO	EECOEECO ⎭	
2.58		CH₂CH₂CH₂CH₂**COCH₂CH₂**COCH₂CH₂CH₂CH₂CH₂	EECOECOEE ⎫	
2.63		COCH₂CH₂**COCH₂CH₂**COCH₂CH₂CH₂CH₂	COECOECOEE ⎬	1,4-Dione
		COCH₂CH₂**COCH₂CH₂**COCH₂CH₂CO	COECOECOECO ⎭	

[a] Proton chemical shifts (ppm) were measured in CDCl₃ at 60° with respect to internal HMDS.
[b] Protons of interest are boldface.

compositions. However, careful examination of the 220 MHz spectrum revealed that the α methylene multiplets consist of two superimposing triplets assignable to hexad sequences of COECOEEE (EECOEECO) and EECOEEE at low and high field, respectively.

In Fig. 2 are presented the 220 MHz $CDCl_3$ solution spectra of several E–CO copolymers. The intensities of the various methylene groups change systematically with copolymer composition. At high CO content the methylene proton resonances of the 1,4-dione structures exhibit additional features which can be attributed to longer sequence placement effects. The high field singlet (at 2.58 ppm) is due to the magnetically equivalent protons (boldface) in

$$CH_2CH_2CH_2CH_2COCH_2CH_2COCH_2CH_2CH_2CH_2,$$

while the boldface methylene groups in

$$COCH_2CH_2COCH_2CH_2COCH_2CH_2CH_2CH_2$$

are no longer equivalent and give rise to an AA'BB' multiplet spectrum centered at about 2.63 ppm. The singlet resonance due to the boldface methylene groups of

$$COCH_2CH_2COCH_2CH_2COCH_2CH_2CO$$

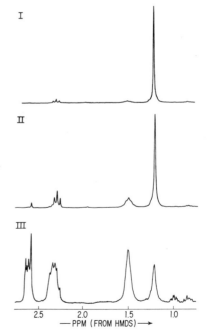

FIG. 2   220 MHz proton spectra of E–CO copolymers in $CDCl_3$ solution. The mol % of CO: I = 7.0; II = 15.7; III = 34.8.

TABLE II

Observed and Calculated Carbon Chemical Shifts of Model Ketones

Compound	Carbon atom	Chemical shift	
		Observed	Calculated
Acetone[a]	1	30.6	31.2
2-Butanone[a]	1	29.0	28.8
	3	36.5	36.5
	4	8.0	7.4
3-Pentanone[a]	1	7.9	7.9
	2	35.4	34.9
2-Heptanone[a]	1	29.6	29.5
	3	43.7	43.7
	4	23.6	23.4
	5	31.5	32.0
3-Heptanone[a]	1	7.8	8.1
	2	35.8	35.3
	4	42.1	41.7
	5	26.2	26.4
4-Heptanone[a]	2	17.4	17.0
	3	44.7	44.7
2-Hexanone[b]	1	29.7	29.5
	3	43.4	43.4
	4	26.2	26.1
2,5-Hexanedione[b]	1	29.7	29.5
	3	37.0	37.0
6-Undecanone[b]	4	23.7	24.1
	5	42.9	42.7
3-Undecanone[a]	1	7.8	8.2
	2	35.5	35.4
	4	42.1	42.4
	5	24.2	24.3
5-Nonanone[b]	3	26.2	26.4
	4	42.6	42.1
3-Methyl butanone[a]	1	27.3	27.3
	3	41.5	41.6
	4	18.1	17.8
2,4-Dimethyl-3-pentanone[c]	1	18.1	18.6
	2	38.3	37.6
2,2-Dimethyl-3-pentanone[c]	1	25.5	25.5
	2	42.7	42.3
2,2,4-Trimethyl-3-pentanone[c]	1	25.1	25.8
	2	43.5	42.6
2,2,4,4-Tetramethyl-3-pentanone[c]	1	28.1	27.7
	2	44.8	45.1
2,6-Dimethyl-4-heptanone[a]	2	24.4	23.9
	3	52.3	52.3

[a] Observed chemical shifts were taken from [10].
[b] Observed chemical shifts were measured in this work.
[c] Observed chemical shifts were taken from [9].

should also appear in this region. In the range of 0.8–1.0 ppm we observed two types of methyl multiplets corresponding to those of ethyl ketones and of alkyl chains containing four or more carbon atoms, from low to high field.

B. Carbon Spectra

1. Model Compounds

Because of the good resolution of carbon-13 NMR and the high sensitivity of the Fourier transform technique, a large number of resonance lines are usually observed in the polymer spectra. In order to facilitate spectral assignments a useful approach is to employ additivity relationships such as those developed by Grant and Paul [8] with the appropriate shielding parameters. This has been demonstrated in numerous applications including our previous studies on poly(vinyl alcohol) [7] and ethylene–vinyl acetate (E–VA) copolymers [6]. For this purpose, we collected published carbon chemical shift data for aliphatic ketones [9,10] and also obtained some results of ketones and diones. From the model compound data in Table II we found that the shielding effects of a carbonyl group on its α or β carbon are not independent of the substituent on that carbon. In fact, three separate substituent parameters have to be defined to account for the α carbon shielding. Similarly, two parameters are needed to precisely describe the shifts due to a β carbonyl group. Table III summarizes the shielding parameters for the carbonyl groups. With these shielding parameters for the carbonyl groups and the alkane parameters of Lindeman and Adams [11] the

TABLE III

Carbon-13 Chemical Shift Substituent Parameters for the Carbonyl Group

Parameter	Number of observations	Mean value	Standard deviation	Description
B_1	5	15.78	.33	Methyl in 2-one
B_2	4	11.68	.05	Methylene, methine or quaternary β shift in 2-one
B'_2	12	12.73	.34	Methylene, methine or quaternary β shift in 3-one or above
C	11	−5.65	.34	γ shift for straight cnain methyl or methylene
C′	6	−4.10	.50	γ shift for methine quaternary or branch methyl

carbon shifts of the model compounds were calculated and compared with the shifts of the model compounds. For the sake of brevity, only those carbon atoms one or two bonds away from the carbonyl groups have been included in one or two bonds away from the carbonyl groups have been included in Table II. The standard deviation of such comparisons is about 0.3 ppm. In Table IV are presented the carbonyl carbon chemical shifts of the model ketones. The shieldings of these carbon atoms exhibit very little regularity. Jackman and Kelly reported that the carbonyl carbon shifts are sensitive to the number of α protons, or, in other words, to the degree of α branching [9]. The data presented in this table generally confirm their observation.

TABLE IV

Carbon-13 Chemical Shifts of Carbonyl Carbon Atoms in Aliphatic Ketones

Number of α protons	Compound	$\delta (C = O)^a$
6	Acetone	206.0
5	2,5-Hexanedione	206.8
	2-Butanone	207.6
	2-Hexanone	208.2
	2-Heptanone	208.4
4	3-Undecanone	209.0
	2,6-Dimethyl-4-heptanone	210.0
	6-Undecanone	210.5
	4-Undecanone	210.6
	5-Nonanone	211.1
	3-Heptanone	211.2
	3-Pentanone	211.4
	3-Methyl-2-butanone	211.8
2	2,2-Dimethyl-3-pentanone	213.4
	2,4-Dimethyl-3-pentanone	215.4
1	2,2,4-Trimethyl-3-pentanone	217.4
0	2,2,4,4-Tetramethyl-3-pentanone	215.4

^{a} Observed carbon chemical shifts were taken from [9].

2. Polymer Spectra

Carbon-13 spectra of the E–CO copolymers were obtained from 10% solutions in 1,2-dideuterotetrachloroethane at 27°. The high field regions are shown in Fig. 3. In general, four main features are observed and their areas change as the copolymer composition is varied. The line at 29.6 ppm increases in intensity and becomes the dominant peak of the spectrum as the CO content decreases. This is also the position expected for a methylene group in a long polymethylene chain. Since the area of the peak at 42.5 ppm

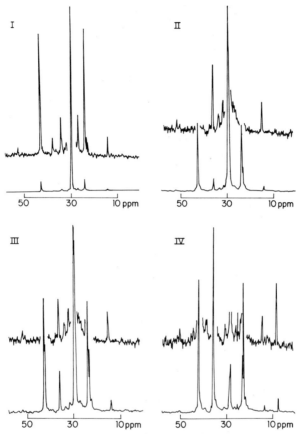

FIG. 3   22.6 MHz carbon-13 spectra of methyl and methylene carbons of E–CO co-polymers in $CDCl_2CDCl_2$ solution. The mol % of CO: I = 7.0; II = 15.7; III = 18.0; IV = 34.8 (peak position in ppm with respect to TMS).

is equal to that at 23.8 ppm for all the copolymers, these features are assigned to the methylene groups α and β to the CO, respectively. The peak at 37.3 ppm is assigned to the α methylene carbons in 1,4-dione structures.

On this basis, the four major methylene resonances can be readily assigned to the three E-centered triads shown in Table V. The α, β and inner methylene lines exhibit fine structure which changes in relative intensities with co-polymer composition. This is probably due to pentad effects. The 1,4-dione methylene line does not show any fine structure, consistent with the fact that the COECO triad gives rise to only a single ECOECOE pentad.

In Table V the predicted chemical shifts for the four major resonances are in good agreement with the observed values. A number of weaker peaks

TABLE V

Carbon-13 Chemical Shifts of Methylene Carbons in Ethylene–Carbon Monoxide Copolymers

		Carbon-13 chemical shift[b]		
Type of carbon	Structure[a]	Predicted	Observed	Triads
α methylene	—COCH$_2$CH$_2$CH$_2$CH$_2$—	42.7	42.5	COEE
1,4-Dione	—COCH$_2$CH$_2$CO—	37.0	37.3	COECO
inner methylene	—CH$_2$CH$_2$CH$_2$CH$_2$CH$_2$CH$_2$—	30.0	29.6	EEE
β methylene	—COCH$_2$CH$_2$CH$_2$CH$_2$—	24.3	23.8	COEE

[a] The carbon of interest is boldface.

[b] Carbon chemical shifts (ppm) were measured in 1,2-dideutero-tetrachloroethane at 27°C with respect to internal TMS.

are also present in the spectra and these are seen more readily in the expanded version (see Fig. 3). They can be assigned to carbons in end groups or branched structures and will be discussed in a later section.

Figure 4 depicts the low field carbonyl resonance of the four E–CO copolymers. They are quite complex and obviously reflect comonomer sequencing. As the CO content is increased intensity moves into the high

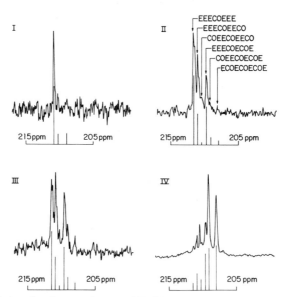

FIG. 4 Carbonyl carbon resonances of E–CO copolymers. The mol % of CO: I = 7.0; II = 15.7; III = 18.0; IV = 34.8 (peak positions in ppm with respect to TMS). The stick plots represent the calculated spectra.

TABLE VI

Carbon Monoxide Centered Sequences

Triads	Pentads	Heptads	Carbon-13 chemical shift[a]
	—COECOECO—	—ECOECOECOE—	208.0
		— COEECOECOE—	209.1
—ECOE—	—EECOECO—		
		—EEECOECOE—	209.7
		—COEECOEECO—	210.4
	—EECOEE—	—EEECOEECO—	211.0
		—EEECOEEE—	211.6

[a] Observed carbonyl carbon chemical shifts were measured in ppm with respect to TMS.

field lines. Assuming an alternating structure, there is only one possible CO-centered triad, three distinguishable pentads, and six distinguishable heptads. These are listed in Table VI. Our assignments were made on the basis of the following considerations. (1) increasing proximity of the carbonyl groups causes an upfield shift, i.e., neighboring γ and ε carbonyl groups contribute -1.8 ppm and $-.6$ ppm, respectively, to the observed carbon shielding; (2) the systematic variation of the relative intensities of the carbonyl resonance lines with copolymer composition. Further support of these assignments was obtained by comparing the observed and calculated spectra. This will be described in Section III.D.

C. Analysis of Copolymer Composition

The simplest quantitative application of NMR data is the determination of copolymer composition. In this study we adopted the method of Ovenall *et al.* developed for E–SO$_2$ copolymers [12]. The average block length is defined as

$$L_{AV} = [\text{Total}]/([1:4] + [\alpha])$$

where [Total] represents the total peak area of all proton or methylene carbon lines; [1:4] the peak area of the proton or methylene lines in 1,4-dione structures, and [α] is the peak area of the protons or methylene carbons α to a carbonyl group. Since the copolymer has the average formula of

$$[CO(CH_2CH_2)_{L_{AV}}]_n$$

the mole percent (and also weight percent since the molecular weights of E and CO are nearly equal) of CO is then given by

$$\%CO = 100/(1 + L_{AV}) \tag{1}$$

TABLE VII

Composition Determination (%CO) of Ethylene–Carbon Monoxide Copolymers

Elemental analysis	Proton NMR	Carbon-13 NMR
7.0	6.2	7.0
16.2	15.8	15.7
18.8	18.3	18.0
22.0	21.8	
33.0	33.7	34.8

Using this formula, copolymer compositions were determined from the proton and carbon-13 spectra. In Table VII the results obtained from the NMR spectra are in good agreement with those determined by elemental analyses.

It should be noted that the peak areas in proton noise-decoupled carbon-13 spectra obtained by the Fourier transform technique are generally not proportional to the number of carbon atoms involved. However, Schaefer and Natusch [13] demonstrated that for many synthetic polymers in solution the nuclear Overhauser effects and spin-lattice relaxation times of carbon atoms in or near the main chain are similar. In such cases, relative peak areas can be used for quantitative analysis. For this reason, we limited our determinations of copolymer compositions and the subsequent monomer sequence analysis to comparison of peak areas from carbons of a given type.

D. Comonomer Sequence Distribution in E–CO Polymers

In principle, comonomer sequence distributions can be derived from copolymerization kinetics. Quantitative calculation of monomer sequence placement is frequently obtainable by using the relative reactivity ratios and feed composition of the comonomers. On the other hand, some insight into the mechanism of copolymerization can also be gained from the experimentally determined sequence distribution of monomers without detailed knowledge of the copolymerization. For the latter case, the observed comonomer sequencing is often compared with the three reference cases: blocky, statistically random, and alternating distributions. In so doing, a qualitative comparison of relative reactivities of monomers in copolymerization can usually be made.

Unlike a completely alternating copolymer, only the CO units in E–CO polymers are isolated by the neighboring E units. Therefore, it is of interest to determine how the E units are distributed among the various sequences. For this purpose, the fraction of E units in 1,4-dione sequences or COECO triads serves as a measure of the ethylene distribution. In the Appendix, sequence distributions in E–CO polymers are considered. For the simplest case or "Bernoullian" distribution, the percent of E units in COECO sequences is given by

$$\%(1,4\text{-dione}) = 100[\%CO/\%E]^2 \qquad (2)$$

where $\%(1,4\text{-dione})$ is the $\%E$ in 1,4-dione structure.

In Table VIII are summarized the values of $\%(1,4\text{-dione})$ obtained from proton and carbon-13 measurements, and the corresponding values are calculated using Eq. (2). The fact that the two sets of the NMR results are in good agreement illustrates the reliability of the NMR methods for determining the monomer sequencing of E–CO copolymers. It is noteworthy that the observed $\%(1,4\text{-dione})$ values for all the copolymers are very close to, but generally lower than, the values calculated from the "random" model. This means that during polymerization of E and CO, addition reaction (3) is somewhat preferred to addition reaction (4):

$$\sim ECOE\cdot + E \longrightarrow \sim ECOEE \qquad (3)$$
$$\sim ECOE\cdot + CO \longrightarrow \sim ECOECO \qquad (4)$$

Using the "random" model, the relative number of CO-centered sequences can be calculated up to heptads and these values are shown as the stick plots in Fig. 4. The calculated relative intensities of the CO-centered heptads follow the observed resonances quite well and our spectral assignments appear to be consistent. However, there are some discrepancies between the observed and calculated carbonyl spectra. For example, in IV of Fig. 4, the relative intensities of the observed ECOECOECOE, COEECOECOE, and EEECOECOE resonances do not match the calculated values, indicating that the "random" model is not followed exactly.

TABLE VIII

Monomer Sequence Distribution ($\%(1,4\text{-Dione})$) in Ethylene–Carbon Monoxide Copolymers

$\%CO^a$	Calculated[b]	Proton NMR[c]	Carbon-13 NMR[c]
7.0	.57	0.7	0.5
15.7	3.5	2.1	2.0
18.0	4.8	4.2	4.1
21.8	7.8	6.0	
34.8	28.4	24.2	27.2

[a] Mole $\%$ of CO in copolymer.
[b] Calculated values of $\%(1,4\text{-dione})$ using Eq. (2).
[c] Observed values of $\%(1,4\text{-dione})$ from the NMR data.

E. End Groups and Branching

If a growing E–CO polymer chain, ending in an E unit, terminates by abstraction of hydrogen from a transfer agent or from another polymer molecule, a methyl end group is formed.

$$RXCH_2CH_2^. + R'H \longrightarrow RXCH_2CH_3 + R'^.$$

When the penultimate unit (X) in the polymer chain is CO, the result is an ethyl ketone. If an E is the penultimate unit ($X = CH_2CH_2$), a butyl end group is formed. For the hypothetical case of a linear, strictly alternating E–CO copolymer with a number-average molecular weight of 11,200 (400 monomer units) in which each chain has two methyl end groups, there would be 0.5 methyl end groups per 100 chain methylene groups. Since the E–CO copolymers which we studied had higher number-average molecular weight values and lower CO contents than this, they must contain fewer methyl end groups. Integration of the region from 0.75–1.00 ppm in the proton spectrum of the copolymer containing 34% of CO and comparison with the methylene proton area show that it contains about 3.6 methyl groups per 100 chain methylene groups. This value far exceeds the maximum possible value for methyl groups and the excess must arise from branching.

Branching in polyethylene has been extensively studied. The well-known "backbiting" mechanism for the formation of short chain branching in the high pressure polymerization of ethylene was proposed by Roedel [14].

(I)

In this mechanism, hydrogen abstraction by a growing polymer radical from the fifth carbon atom in the chain results in a butyl branch. Convincing evidence for such short chain branching comes from recent carbon-13 NMR studies of commercial high-pressure polyethylenes by Randall [15] and by Bovey and coworkers [16]. Both concluded that the principle type of short chain branching present was butyl with indications of much smaller amounts of ethyl branches.

The Roedel mechanism can be readily extended to E–CO copolymers. Hydrogen abstraction from carbon atoms adjacent to carbonyl groups should be energetically favored and a number of possible mechanisms can be envisioned.

$$R-\underset{\underset{\displaystyle CH_2}{\overset{\displaystyle |}{\underset{\displaystyle CO}{|}}}}{\overset{\displaystyle H-\!-\!-\dot{C}H_2}{\underset{\displaystyle \diagdown\;\diagup}{CH}}}\;\underset{\displaystyle CH_2}{\overset{\displaystyle |}{\underset{\displaystyle \diagdown}{CH_2}}} \longrightarrow R\dot{C}HCOCH_2CH_2CH_2CH_3 \xrightarrow{CH_2=CH_2} R-\underset{\underset{\displaystyle CH_2}{\overset{\displaystyle |}{\underset{\displaystyle CH_2}{\dot{C}H_2}}}}{\overset{\displaystyle \dot{C}H_2}{CH}}COCH_2CH_2CH_2CH_3$$

(II)

Mechanism (II) would give rise to a butyl ketone branch. Backbiting by a radical ending in —COCH$_2$CH$_2$ would give rise to a branch ending in an ethyl ketone group

$$\underset{\underset{\displaystyle CO}{\underset{\displaystyle \diagup}{\overset{\displaystyle CH_2}{|}}}}{\overset{\displaystyle H}{\underset{\displaystyle \diagdown}{RCH}}}\quad\underset{\underset{\displaystyle CH_2}{\overset{\displaystyle |}{\underset{\displaystyle \diagdown}{\dot{C}H_2}}}}{} \longrightarrow R\dot{C}HCH_2COCH_2CH_3 \xrightarrow{CH_2=CH_2} R\underset{\underset{\displaystyle CH_2}{\overset{\displaystyle |}{\underset{\displaystyle CH_2}{\dot{C}H_2}}}}{\overset{\displaystyle \dot{C}H_2}{C}}HCH_2COCH_2CH_3$$

(III)

In the methyl proton region of Fig. 5 the resonances at higher field consist of two triplets center at 0.809 ppm and 0.836 ppm (with a $J = 7$ Hz) and

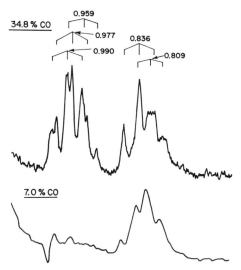

FIG. 5.   220 MHz proton spectra of the methyl proton region of E–CO copolymers. The spectra of 34.8% and 7.0% were obtained via continuous wave and Fourier transform modes, respectively. (The chemical shifts are given in ppm with respect to TMS.)

TABLE IX

Short Chain Branching
in Ethylene–Carbon
Monoxide Copolymerization

| %CO | Methyl groups per 100 chain methylene groups | |
	Ethyl ketone	n-Butyl
7.0	~0	1.9
15.7	0.8	1.5
21.8	1.3	2.1
34.8	1.8	1.8

can be assigned to two methyl groups in butyl branches as described in (I) and (II). Similarly, at least three overlapping triplets are recognizable in the lower field peaks. These triplets are attributable to ethyl ketone branching as in (III) and their chemical shift differences probably arise from the different penultimate comonomeric units in the polymer chain. Quantitative measurements of the methyl groups by spectral integration are summarized in Table IX. For all the copolymers the methyl groups characteristic of butyl branching remain reasonably constant. The average value of 1.8 groups per 100 chain methylene groups is very similar to those observed in the high pressure polyethylene [16]. On the other hand, the ethyl ketone content was found to increase monotonically with increasing CO content in the copolymer. The carbon-13 chemical shifts for the branched structures were calculated using the Lindeman and Adams formula [11] together with the ketone parameters. The predicted values are given in Table X. The butyl or ethyl ketone methyl carbons from both end groups and branch structures have characteristic predicted chemical shifts of 13.9 ppm and 8.2 ppm. Weak peaks at 13.8 ppm are visible in the carbon-13 spectra of the copolymers (Fig. 3), confirming the presence of butyl methyls. In the spectrum of the polymer with 34% CO a peak appears at 7.6 ppm, which is assigned to ethyl ketone methyls. This peak is not seen or is much weaker in the spectra of the copolymers containing less CO consistent with our observation that the number of ethyl ketone methyls increases relative to the number of butyl methyls as the CO content of the copolymer is increased.

In all the carbon-13 spectra weak peaks are seen in the region 50–52 ppm. These are assigned to methine carbon atoms in branches originating next to carbonyl groups. The predicted chemical shift for such a carbon in one possible branched structure given in Table X is 49.8 ppm. Several additional

TABLE X

Predicted Carbon-13 Chemical Shifts of End Groups and Branched Structures in Ethylene–Carbon Monoxide Copolymers

I. *End groups*

$$R-CH_2-CH_2-CO-CH_2-CH_3$$
$$\quad\; 24.3 \quad 42.4 \qquad\quad 35.4 \quad\; 8.2$$

$$R-CO-CH_2-CH_2-CH_2-CH_3$$
$$\qquad\quad 42.4 \quad 26.8 \quad 22.7 \quad 13.9$$

$$R-CH_2-CH_2-CH_2-CH_3$$
$$\quad\; 29.7 \quad 32.4 \quad 22.7 \quad 13.9$$

II. *Branched structures*

$$\overset{\displaystyle 34.2 \quad 30.0 \quad\; 22.9 \quad\; 13.9}{CH_2-CH_2-CH_2-CH_3}$$
$$|$$
$$\underset{\displaystyle 30.2 \quad\; 27.5 \quad\; 34.5 \quad\; 37.1 \quad 34.5 \quad 27.5 \quad 30.2}{-CH_2-CH_2-CH_2-CH-CH_2-CH_2-CH_2-}$$

$$\overset{\displaystyle 40.0 \quad\; 27.0 \quad\; 22.7 \quad\; 13.9}{CO-CH_2-CH_2-CH_2-CH_3}$$
$$|$$
$$\underset{\displaystyle 30.2 \quad\; 27.5 \quad\; 28.8 \quad\; 49.8 \quad 28.8 \quad 27.5 \quad 30.2}{-CH_2-CH_2-CH_2-CH-CH_2-CH_2-CH_2-}$$

$$\overset{\displaystyle 47.0 \qquad\qquad\quad\; 35.6 \quad\; 8.2}{CH_2-CO-CH_2-CH_3}$$
$$|$$
$$\underset{\displaystyle 30.2 \quad\; 27.5 \quad\; 34.5 \quad\; 31.4 \quad 34.5 \quad 27.5 \quad 30.2}{-CH_2-CH_2-CH_2-CH-CH_2-CH_2-CH_3}$$

a Chemical shift values are given directly above or below the carbon of interest.

weak peaks are visible in the spectra between the main methylene carbon peaks. In the sample containing the least amount of CO, these peaks can be identified with carbon atoms in the normal Roedel branched structure (Table X).

In contrast to the good agreement between the proton and carbon-13 data for the copolymer compositions, the carbon-13 spectra gave much lower methyl group concentrations than the proton spectra. This discrepancy is almost certainly due to the fact that the spin-lattice relaxation times of the methyl group carbons are much longer than those of the methylene carbons. This discrepancy was reduced, but still present, when the time between pulses was increased from 0.7 seconds to 2.0 seconds. In view of these difficulties, we have not used the carbon-13 technique for quantitative determination of branching in E–CO polymerization.

F. Conformational Transitions of Polyethylene Segments

Our previous studies of solvent effects on the proton spectra of ethylene copolymers revealed that the spectral features of the methylene protons of the polyethylene segments are strongly dependent on the nature of the solvent. Unlike the singlet resonance peak in the spectra obtained in aliphatic chlorohydrocarbon solution, these alkane-like methylene protons give rise to two peaks in the 220 MHz spectra [17] in aromatic solvents. Because of the similarity of these features to those observed for polyethylene oligomers [18], we assigned the doubled alkane-like methylene proton peaks to the two types of intramolecular structures, "polymeric" and "monomeric" species of the ethylene sequences. The "polymeric" species probably arises from intramolecular chain folding of the polyethylene segments containing eight or more ethylene units. Moreover, the existence of polymeric structure

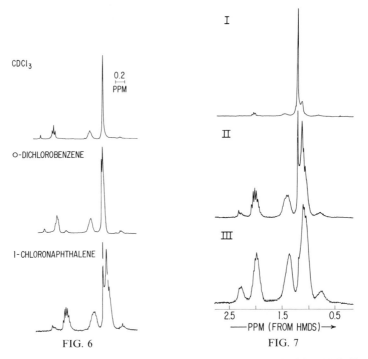

FIG. 6   FIG. 7

FIG. 6   220 MHz proton spectra of an E–CO copolymer (mol % CO = 15.7) dissolved in several solvents.

FIG. 7   220 MHz 1-chloronaphthalene solution spectra of several E–CO copolymers. The mol % of CO: I = 7.0; II = 15.7; III = 22.0.

in these ethylene copolymers appears to depend on their compositions and, to a lesser extent, on monomer sequence distribution [19].

Figure 6 depicts the effects of solvents on the methylene proton spectra of an E–CO copolymer. In aromatic solvents such as *ortho*-dichlorobenzene and 1-chloronaphthalene the resonances due to the alkane-like methylene protons at 1.2 ppm are also split into two lines. Since these spectra are remarkably similar to those of our previously studied ethylene copolymers including E–VA, ethylene–vinyl formate, and ethylene–vinyl chloride polymers [17,19] we again assigned the sharp line at the lower field to the equivalent internal methylene groups in the "polymeric" polyethylene segments and the broad upfield multiplets to the disordered "monomeric" methylene groups.

The effects of copolymer composition on the intramolecular structure formation are illustrated in Fig. 7. As in the cases of other ethylene copolymers, the spectra of the alkane-like methylene protons are indeed dependent on the comonomer (CO) contents in the polymers. The lower field "polymeric" singlet decreases rapidly in intensity with increasing CO content.

In comparison with a statistically random ethylene copolymer, the probability of finding E units in the long ethylene sequences of an alternating ethylene copolymer should be considerably smaller. This point is clearly illustrated in Fig. 8, in which G(8), the observed fraction of E units in

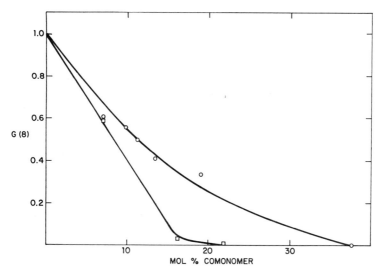

FIG. 8   Effects of comonomer sequence distribution on intramolecular conformational structure of ethylene copolymers. ○, random E–VA copolymers; □, alternating E–CO polymers.

"polymeric" species containing eight or more units, are plotted against the comonomer contents for E–VA and E–CO polymers. Unlike the statistically random E–VA polymers, G(8) of E–CO polymers decreases sharply with increasing comonomer content and becomes negligible when it reaches 22%. On the other hand, the value of G(8) of E–VA polymers does not become zero until the VA content exceeds 37.5%.

APPENDIX. SEQUENCE DISTRIBUTIONS IN SEMIALTERNATING COPOLYMERS

In a completely alternating copolymer there is zero probability that two successive monomer units will be of the same type and the copolymer has 1:1 molar composition. We have chosen to use the term "semialternating copolymers" for those in which single units of one comonomer are isolated between runs or blocks of one or more units of the second comonomer. In cases, such as the present, where tacticity need not be considered, the problem is one of determining how these comonomer units are distributed among the blocks.

Consider a semialternating copolymer containing %E ethylene and %CO carbon monoxide. For every 100 monomer units there are %CO runs of carbon monoxide, each run containing one carbon monoxide. There are also %CO runs of ethylene containing an average of (%E)/(%CO) ethylenes per run. Each ethylene run must contain at least one ethylene, leaving (%E − %CO) ethylenes to be distributed among the %CO runs.

If we assume a Bernoullian model in which the probability of finding an E next to an E is fixed, the relative numbers of ethylene blocks of different lengths can be readily calculated. Let P_E be the probability that a given E is adjacent to a second E. Then $(1 - P_E)$ is the probability that a given E is adjacent to a CO. Since each ethylene run begins with COE, the probability of a run of n ethylenes, P_n is given by:

$$P_n = P_E^{n-1}(1 - P_E)$$

For a random distribution of the (%E − %CO) ethylenes among the %CO runs;

$$P_E/(1 - P_E) = (\%E - \%CO)/(\%CO)$$

giving

$$P_E = 1 - (\%CO)/(\%E)$$

Note that when %E = %CO = 50, $P_E = 0$, as required for a completely alternating copolymer. When %E goes to 100, P_E goes to unity, giving longer and longer blocks.

The fractional number of E runs which contain one E is given by:

$$P_1 = (1 - P_E) = (\%CO)/(\%E)$$

Since the average length of an E run is $(\%E)/(\%CO)$, the percentage of E units which are in COECO or 1:1 sequences, $\%(1,4\text{-dione})$, is given by:

$$\%(1,4\text{-dione}) = 100(\%CO)^2/(\%E)^2$$

The probabilities of the six possible carbon monoxide centered heptads can be calculated from this model, bearing in mind that each carbon monoxide unit must be flanked by ethylenes. The expressions are given in Table XI.

TABLE XI

Probabilities of Carbon Monoxide Centered
Heptads Given by Bernouillian Model

Heptad	Probability
ECOECOECOE	$(1 - P_E)^2$
COEECOECOE	$2P_E(1 - P_E)^2$
EEECOECOE	$2P_E^2(1 - P_E)$
COEECOEECO	$P_E^2(1 - P_E)^2$
EEECOEECO	$2P_E^3(1 - P_E)$
EEECOEEE	P_E^4
$P_E = 1 - (\%CO)/(\%E)$	

ACKNOWLEDGMENTS

The authors would like to acknowledge Drs. C. F. Hammer and J. M. Edinger for supplying the ethylene–carbon monoxide copolymers and some characterization data. Special thanks are due to R. O. Balback, F. W. Barney, Jr., and G. Watunya for technical assistance. Moreover, it is the authors' pleasure to thank Drs. G. Pruckmayr and M. L. Sheer for reviewing the manuscript.

REFERENCES

1. T. K. Wu, *J. Polym. Sci. Part A-2* **14**, 343 (1976).
2. M. M. Brubaker, D. D. Coffman, and H. H. Hoehn, *J. Amer. Chem. Soc.* **74**, 1509 (1952).
3. Y. Morishima, T. Takizawa, and S. Murahashi, *Eur. Polym. J.* **9**, 669 (1973).
4. G. C. Alfonso, L. Fiorina, E. Martuscelli, E. Pedemonte, and S. Russo, *Polymer* **14**, 373 (1973). See also the references therein.
5. T. K. Wu, *J. Phys. Chem.* **73**, 1801 (1969).
6. T. K. Wu, D. W. Ovenall, and G. S. Reddy, *J. Polym. Sci., Part A-2* **12**, 901 (1974).
7. T. K. Wu and D. W. Ovenall, *Macromolecules* **6**, 582, (1973).
8. D. M. Grant and E. G. Paul, *J. Amer. Chem. Soc.* **86**, 2984 (1964).
9. L. M. Jackman and D. P. Kelly, *J. Chem. Soc. B*, 102 (1970).

10. L. F. Johnson and W. C. Jankowski, "Carbon-13 NMR Spectra." Wiley (Interscience), New York, 1972.

11. L. P. Lindeman and J. Q. Adams, *Anal. Chem.* **43**, 1245 (1971).

12. D. W. Ovenall, R. S. Sudol, and G. A. Cabat, *J. Polym. Sci. Part A-1* **11**, 233 (1973).

13. J. Schaefer and D. F. S. Natusch, *Macromolecules* **5**, 416 (1972).

14. M. J. Roedel, *J. Amer. Chem. Soc.* **75**, 6110 (1953).

15. J. C. Randall, *J. Polym. Sci. Part A-2* **11**, 275 (1973).

16. D. E. Dorman, E. P. Otocka, and F. A. Bovey, *Macromolecules* **5**, 574 (1972).

17. T. K. Wu, *Macromolecules* **2**, 520 (1969); **3**, 610 (1970).

18. K. J. Liu, *J. Polym. Sci. Part A-2* **4**, 155 (1966); **5**, 1209 (1967); **6**, 947 (1968).

19. T. K. Wu, *Macromolecules* **6**, 737 (1973).

4

Instrumental Characterization of Ethylene–Ethyl Acrylate–Carbon Monoxide Terpolymers

J. E. McGRATH

CHEMISTRY DEPARTMENT
VIRGINIA POLYTECHNIC INSTITUTE
    AND STATE UNIVERSITY
BLACKSBURG, VIRGINIA

L. M. ROBESON   *T. E. NOWLIN*
B. L. JOESTEN   *E. W. WISE*

RESEARCH AND DEVELOPMENT DEPARTMENT
UNION CARBIDE CORPORATION
BOUND BROOK, NEW JERSEY

I. INTRODUCTION

The preparation and characterization of ethylene copolymers has been discussed extensively in the literature [1–6]. The comonomer generally reduces the degree of crystallinity relative to that of low density polyethylene (LDPE) and may provide enhanced processibility or physical properties [7]. Vinyl acetate, alkyl acrylates, carbon monoxide, and many other co-monomers have been investigated [1,3]. Carbon monoxide is a particularly interesting comonomer [8]. It is readily available, economical, and confers polarity and photodegrading [9–11], properties on polyethylene systems. The kinetics [12], mechanism [13], and morphology [14] of ethylene–carbon monoxide copolymers have been studied in detail.

The resulting copolymers have a polyketonic structure of the type shown in (I), where n changes as a function of the monomer feed

$$\text{wwww}(CH_2-CH_2)_n-(CO)\text{www}$$

(I)

composition. However, many investigators [1,8] have noted that the crystallinity of these copolymers changes very little with carbon monoxide content over a wide compositional range, and it has been reported [15] that they are cocrystalline.

The special effectiveness of ethylene in ethylene–carbon monoxide copolymerizations has been previously reported [8]. It was stated that ethylene forms a 1:1 complex with carbon monoxide. The complex may then "homopolymerize" or "copolymerize" with excess ethylene. Alternatively, it is sometimes possible to incorporate a third monomer into the macromolecule. Carbon monoxide (CO) alone is not capable of polymerization and cannot add to a growing chain having a CO end group [e.g., (II)].

$$\text{wwww}CH_2-CH_2-\overset{\overset{\textstyle O}{\textstyle \|}}{C}$$

(II)

The upper limit for CO incorporation is $50\,mol\%$, which is, of course, also $50\,wt\%$ in the special case of ethylene–carbon monoxide copolymerization.

Terpolymers containing ethylene, carbon monoxide, and a third vinyl monomer [8,16] have been much less extensively investigated. Terpolymers of ethylene, vinyl acetate and carbon monoxide (E/VA/CO) have been discussed in the patent literature [17,18].

It was reasoned that terpolymerization of alkyl acrylates (e.g., ethyl, 2-ethylhexyl) with ethylene and carbon monoxide would be possible and should lead to materials displaying lower crystallinity than either LDPE or the cocrystalline E–CO system. Moreover, the presence of the CO repeat unit was expected to produce a more polar system than the ethylene–ethyl acrylate (E/EA) copolymer. The CO unit should be capable of specific interactions with polymer additives, solvents, and even other polymers. Co- and terpolymers containing alkyl acrylates have advantages in thermal processing stability over those of vinyl acetate. It is well known that ethylene–vinyl acetate copolymers (E/VA) can liberate acetic acid at 200°C or even lower. However, reactivity ratio considerations [1,3] favor the synthesis of more compositionally uniform ethylene copolymers from vinyl acetate than ethyl acrylate.

This chapter discusses the synthesis of ethylene–ethyl acrylate–carbon monoxide (E/EA/CO) terpolymers and their characterization by proton magnetic resonance (PMR), infrared, and differential scanning calorimetry (DSC). A separate report [19] discusses the transitions and relaxational behavior of these terpolymers and their miscibility with polyvinyl chloride.

II. EXPERIMENTAL TECHNIQUE

General techniques utilized for high pressure ethylene co- or terpolymerization have been previously published [1,3,20]. The terpolymerization of high purity ethylene, carbon monoxide, and ethyl acrylate was conducted in a small batch stirred reactor. The ethyl acrylate (Union Carbide Corporation) was freshly distilled prior to polymerization. Polymerizations were conducted in benzene solvent at 160°C and ~15,000 psi with azobisisobutyronitrile (AIBN) initiator. Conversions were limited to ~20%. Compositions were varied by charging different molar ratios of the three monomers. The terpolymers were recovered by coagulation into methanol and vacuum drying to constant weight. A few terpolymers containing 2-ethylhexyl acrylate and some control copolymers were also prepared using similar procedures.

The melt index (190°C, 44 psi) of all samples was in the range 0.1–5.0 gm/10 min. The composition by PMR of the terpolymers prepared during this investigation are shown in Table I.

100-MHz PMR spectra of an ethylene–ethyl acrylate (EEA) copolymer, the ethylene–ethyl acrylate–carbon monoxide (E/EA/CO) terpolymers,

TABLE I

Composition Values for Ethylene–Ethyl Acrylate–Carbon Monoxide Terpolymers

Terpolymer number	Mole percent			Weight percent		
	Ethylene	Ethyl acrylate	Carbon monoxide	Ethylene	Ethyl acrylate	Carbon monoxide
E/EA control	92.2	7.8	—	77	23	—
E/CO control	81	—	19	81	—	19
E/VA/CO control – 1	—	—	—	66	23[b]	10.5
– 2	—	—	—	56	32[b]	10.2
E/EA/CO control – 1	77.6	3.2	19.2	71.8	10.5	17.7
– 2	73.7	6.3	19.9	63.6	19.4	17.2
– 3	83.2	2.2	14.6	78.8	7.4	13.8
– 4	82.6	1.5	15.9	79.6	5.1	15.3
– 5	79.9	1.6	18.5	76.8	5.5	17.8
– 6	67.6	8.5	23.7	55.6	24.8	19.6
– 7	64.0	13.3	23.6	47.8	35.4	16.9
– 8	53.7	22.2	24.1	34.2	50.5	15.3
– 9	86.6	8.5	4.9	70.9	25.1	4.0
– 10	81.8	8.8	9.4	66.7	25.7	7.7
– 11	—	—	—	62.1	28.4	9.4
– 12	—	—	—	63	28.7	8.3
– 13	—	—	—	64.3	26.7	8.5
– 14	—	—	—	64.9	26.1	9.0
– 15	—	—	—	66.1	26.3	7.6
– 16[a]	67.6	7.4	25.1	47.9	34.2	17.8
– 17	62.9	9.1	28.0	41.8	39.6	18.6

[a] 2-Ethyl hexyl acrylate.
[b] Vinyl acetate.

and the ethylene–2-ethyl-hexyl acrylate–carbon monoxide (E/2EHA/CO) terpolymers were recorded in the following manner.

A Varian HA-1000-15 NMR spectrometer was operated in the field sweep mode. Spectra were recorded at about 120°C in 10 mm sample tubes. *Para*-dichlorobenzene served as internal standard and lock. Solutions were approximately 8 wt% polymer in tetrachloroethylene.

High EA content materials were not soluble in tetrachloroethylene. However, these samples were soluble in CDCl₃ at ambient temperatures. These solutions were placed in a 5-mm tube and tetramethylsilane (TMS) was used as lock and internal standard. Standard deviations indicate the precision of these measurements was ±1–2%.

III. COMPOSITIONAL ANALYSIS AND SEQUENCE DISTRIBUTION BY PROTON MAGNETIC RESONANCE (PMR)

The spectrum of each type of polymer examined will be discussed separately.

A. Ethylene–Ethyl Acrylate (EEA) Copolymer

The spectrum of an EEA copolymer is shown in Fig. 1. Peak assignments and chemical shifts relative to TMS are reported in Table II.

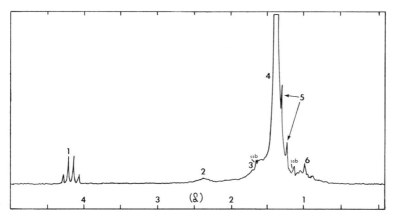

FIG. 1   100-MHz spectrum of an ethylene/ethyl acrylate (EEA) copolymer. EA content calculated to be 14.6 wt %.

TABLE II

PMR Peak Assignments of an Ethylene–Ethyl Acrylate (EEA) Copolymer[a]

Peak[b]	Structure (proton in boldface)	Chemical shift (ppmδ)

1	—CH$_2$—C—**H**—
	   |
	   C=O
	   |
	   O—**CH$_2$CH$_3$**
		4.17

2	—CH$_2$—**CH**—
	   |
	   C=O
	   |
	   O—CH$_2$CH$_3$
		2.38

3	—**CH$_2$**—CH—
	   |
	   C=O
	   |
	   O CH$_2$CH$_3$
		~1.65

4	—CH$_2$CH$_2$—	1.38

5	—CH$_2$—CH—(partially resolved)
	   |
	   C=O
	   |
	   O—CH$_2$**CH$_3$**
		1.30

6	—CH$_2$CH$_2$**CH$_3$**	0.98

[a] The branching methylene protons in EA (peak 1) at 4.17 ppm are shown as a well resolved quartet exhibiting typical proton coupling of about 7 Hz. Peak 2 is assigned to the chain methine proton in EA and is one-half the area of peak 1. The chain methylene protons in EA (peak 3) and the methyl group in EA (peak 5) are only partially resolved due to the ethylene chain protons (peak 4).

[b] See Fig. 1.

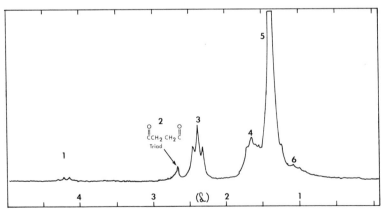

FIG. 2 100-MHz spectrum of an E/EA/CO terpolymer at 120°C. Low EA content (7.4 wt %).

B. Ethylene–Ethyl Acrylate–Carbon Monoxide (E/EA/CO) Terpolymers

Examples of the spectra obtained from the (E/EA/CO) terpolymers examined are shown in Figs. 2–5.

Peak assignments and chemical shifts relative to TMS are shown in Table III. As can be seen by comparing Figs. 2–5 with Fig. 1, the PMR spectrum becomes more complex once carbon monoxide is incorporated into the polymer.

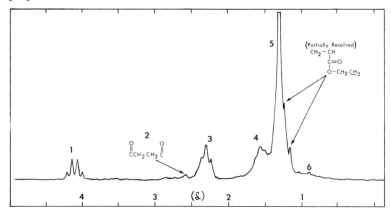

FIG 3   100-MHz spectrum of an E/EA/CO terpolymer at 120°C (7.66 wt % CO). (Note that components distinguished in figures by an underline are distinguished in the text by boldface.)

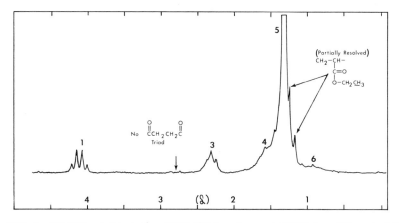

FIG. 4   100-MHz spectrum of an E/EA/CO terpolymer at 120°C. Low CO content (4.0 wt %). Note the absence of the COECO triad.

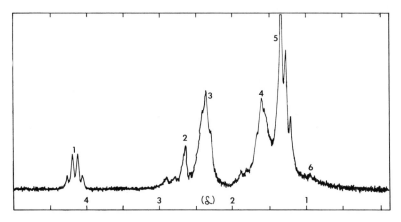

FIG. 5   100-MHz spectrum of an E/EA/CO terpolymer at ambient temperature in CDCl₃. High (35.4 wt %) EA content.

TABLE III

PMR Peak Assignments of Ethylene–Ethyl Acrylate–Carbon Monoxide (E/EA/CO) Terpolymers

Peak[a]	Structure (proton in boldface)	Chemical shift (ppmδ)

1	$-CH_2-CH-$
	         $\overset{\mid}{C}=O$
	         $\overset{\mid}{O}-\mathbf{CH_2CH_3}$
		4.18

2	$\overset{O}{\overset{\|}{-C}}-\mathbf{CH_2CH_2}\overset{O}{\overset{\|}{C}}-$	2.66
3	$\overset{O}{\overset{\|}{-C}}-\mathbf{CH_2CH_2CH_2}-$ and $CH_2-\mathbf{CH}$ $\mid$	2.38
4	$\overset{O}{\overset{\|}{-C}}-\mathbf{CH_2CH_2CH_2}-$ and $\mathbf{CH_2}-CH$ $\mid$	~1.65

5	$CH_2\mathbf{CH_2}CH_2-$ and $\mathbf{CH_2}$ CH
	                               $\overset{\mid}{C}=O$
	                               $\overset{\mid}{O}-CH_2CH_3$
		1.38

6	$-CH_2CH_2\mathbf{CH_3}$	~1.0

	$\overset{O}{\overset{\|}{-C_2}}\mathbf{CH_2CH}CH_2-$
	                $\overset{\mid}{C}=O$
	                $\overset{\mid}{O}-CH_2CH_3$
		~2.7–2.8
		 Tentative assignment

[a] See Figs. 2–5.

Peaks 2, 3, and 4 are due to structures in which carbon monoxide is adjacent to an ethylene unit. Peak 2 is assigned to the triad,

$$\begin{matrix} O & & O \\ \parallel & & \parallel \\ C-CH_2CH_2C \end{matrix}$$

Peaks 3 and 4 are the α and β methylene protons, respectively, adjacent to a carbonyl.

Structures in which the carbonyl is adjacent to the ethyl acrylate, either

$$\begin{matrix} O \\ \parallel \\ -C-CH_2-CH- \\ \quad\quad\quad | \\ \quad\quad\quad C=O \\ \quad\quad\quad | \\ \quad\quad\quad O-CH_2CH_3 \end{matrix} \quad\quad \text{or} \quad\quad \begin{matrix} O \\ \parallel \\ -CH_2-CH-C- \\ \quad\quad | \\ \quad\quad C=O \\ \quad\quad | \\ \quad\quad O-CH_2CH_3 \end{matrix}$$

$$\text{(III)} \quad\quad\quad\quad\quad\quad\quad\quad \text{(IV)}$$

have not been positively identified. However, the effect of placing a carbonyl group adjacent to an ethylene unit can be used to give important information as to the probable *existence* of structure (III) and *nonexistence* of structure (IV). For example, an ethylene block, as in polyethylene, gives a single line at 1.32 ppm. Placing a carbonyl adjacent to one ethylene unit,

$$\begin{matrix} O \\ \parallel \\ C-CH_2CH_2CH_2 \end{matrix}$$

gives a line at 2.38 ppm or a shift downfield of 1.06 ppm. Consequently, a similar effect would be expected in ethyl acrylate. In EA the methylene chain protons, CH_2-CH-, give a peak at about 1.65 ppm. Therefore, one would predict a

$$\begin{matrix} O \\ \parallel \\ CCH_2-CH \\ \quad\quad\quad | \\ \quad\quad\quad C=O \\ \quad\quad\quad | \\ \quad\quad\quad O-R \end{matrix}$$

structure to give a peak at $(1.65 + 1.06) = 2.71$ ppm. In the E/EA/CO samples that contain a significant amount of EA (> 19 wt %), a peak at 2.7–2.9 ppm is observed (see Figs. 3 and 5). This is tentatively assigned to the methylene chain protons in structure (III).

Similar reasoning predicts that structure (IV) should give a methine peak,

$$\begin{matrix} O \\ \parallel \\ CH_2CHC \\ \quad\quad | \\ \quad\quad C=O \\ \quad\quad | \\ \quad\quad O-R \end{matrix}$$

at $(2.38 + 1.06) = 3.44$ ppm. All spectra exhibited no peak near this chemical shift. Consequently, it is assumed that the EA–CO sequence shown as structure (IV) does not exist. This is consistent with the literature [8,16], (e.g., it is believed that carbon monoxide forms a complex with ethylene prior to polymer formation).

A summary of the results of these analyses is given in Table III.

C. Ethylene–2-Ethyl Hexyl Acrylate–Carbon Monoxide (E/2EHA/CO) Terpolymers

The PMR spectra of two E/2EHA/CO samples examined were similar and only one is shown in Fig. 6. Peak assignments and chemical shifts relative to TMS are reported in Table IV.

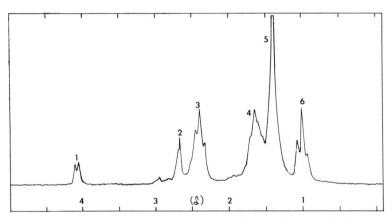

FIG. 6   100-MHz spectrum of an E/2EHA/CO terpolymer at 120°C.

In these polymers the branching ethyl group in EA is replaced with the bulky 2-ethyl hexyl group. Consequently, peak 1 at 4.07 ppm is now a doublet due to coupling with only one proton, and the two terminal methyl groups on the branch give rise to a more intense peak at 1.0 ppm.

In all of the samples studied, the singlet at 2.67 ppm was assigned to the carbon monoxide–ethylene–carbon monoxide triad,

$$\underset{\displaystyle -\!\overset{\displaystyle O}{\overset{\displaystyle \|}{C}}CH_2CH_2\overset{\displaystyle O}{\overset{\displaystyle \|}{C}}\!-}{}$$

This triad was detected in all E/EA/CO and E/2EHA/CO samples that contained greater than about 4 wt % carbon monoxide. Consequently, this

TABLE IV

PMR Peak Assignments of Ethylene–2-Ethyl Hexyl Acrylate–Carbon Monoxide Terpolymers

Peak[a]	Structure (proton in boldface)	Chemical shift (ppmδ)

1	$-CH_2-CH-$
	     │
	     C=O
	     │
	     O
	     │
	     CH₂C—C—C—C—C
	        │
	        C—C
		4.07

2	O O ‖ ‖ $-$C**CH₂CH₂**C$-$	2.66
3	O ‖ $-$C**CH₂CH₂**CH₂$-$ and **CH₂**—CH	2.38
4	O ‖ $-$CCH₂CH₂**CH₂**$-$ and **CH**—**CH**$-$	∼1.65
5	$-$CH₂**CH₂**CH₂$-$	1.38

6	$-CH_2CH-$
	     │
	     C=O
	     │
	     O
	     │
	     C—C—C—C—C—**CH₃** and $-$CH₂CH₂**CH₃**
	        │
	        C—**CH₃**
		1.0

[a] See Fig. 6.

triad was not seen in sample 9 (see Table I and Fig. 4). A similar triad,

$$\underset{\qquad}{\overset{O}{\underset{\|}{C}}}-CH_2-\underset{\underset{\underset{OCH_2CH_3}{|}}{\overset{|}{C=O}}}{CH}-\overset{O}{\underset{\|}{C}}-$$

between ethyl acrylate and carbon monoxide is probably not present in any of the samples. It was estimated that the chain methine proton adjacent to a carbon monoxide should give a peak near 3.44 ppm. No peak was found in this region in any of the spectra. Consequently, it is assumed that this type of triad is not present in any appreciable amounts.

Structures in which one carbon monoxide is adjacent to two other monomer units to give the sequences:

$$\begin{matrix} \text{O} \\ \| \\ \text{C--CH}_2\text{CH}_2\text{--CH}_2\text{CH}_2\text{--} \end{matrix} \qquad \text{(COEE)}$$

$$\begin{matrix} \text{O} \\ \| \\ \text{C--CH}_2\text{CH}_2\text{CH}_2\text{--CH--} \\ \qquad\qquad\qquad | \\ \qquad\qquad\qquad \text{C}{=}\text{O} \\ \qquad\qquad\qquad | \\ \qquad\qquad\qquad \text{O--CH}_2\text{--CH}_3 \end{matrix} \qquad \text{(COEEA)}$$

$$\begin{matrix} \text{O} \\ \| \\ \text{CCH}_2\text{--CH--CH}_2 \qquad \text{CH--} \\ \qquad\quad | \qquad\qquad\quad | \\ \qquad\quad \text{C}{=}\text{O} \qquad\qquad \text{C}{=}\text{O} \\ \qquad\quad | \qquad\qquad\quad | \\ \qquad\quad \text{O} \qquad\qquad\quad \text{O} \\ \qquad\quad | \qquad\qquad\quad | \\ \qquad\quad \text{CH}_2\text{--CH}_3 \quad\ \ \text{CH}_2\text{--CH}_3 \end{matrix} \qquad \text{(COEAEA)}$$

are likely present in both the EA and 2EHA terpolymers.

In addition, the results suggest that structures involving a carbon monoxide moiety adjacent to an ethylene are more favorable than structures which involve a CO adjacent to an ethyl acrylate. The fact that the triad COECO is usually present, whereas the triad COEACO is probably absent, in all samples supports this contention. Moreover, peaks due to ethylene–carbon monoxide sequences are much more intense than the peak assigned to a CO unit adjacent to an EA moiety. See, for example, Fig. 5. Peaks 2, 3, and 4 (E/CO sequences) are much more intense than the peak involving an EA/CO sequence (2.7–2.9 ppm).

IV. ANALYSIS BY INFRARED

The infrared (ir) spectra of the terpolymers examined clearly portray two different carbonyl groups between 5.7 μm and 5.9 μm. The carbonyl from the acrylate moiety is at 5.75 μm and the carbonyl from the chain is at 5.85 μm.

V. ANALYSIS BY DIFFERENTIAL SCANNING CALORIMETRY (DSC)

Three of the E/EA/CO terpolymer samples have been examined by DSC along with two E/VA/CO control resins. The data are shown in Table V. Although exactly the same compositions for the two different vinyl co-

TABLE V

Crystallinity of E/EA/CO and E/VA/CO Terpolymers by Differential Scanning Calorimetry[a]

Terpolymer number	Ethylene (wt %)	Vinyl acetate (wt %)	CO (wt %)	Ethyl acrylate (wt %)	ΔH (cal/gm)	$T_{M_1}^b$
E/VA/CO control 1	66	23	11	—	13.0	59
E/VA/CO control 2	57	32	10	—	6.7	43
E/EA/CO 1	71	—	18	11	20.7	107
E/EA/CO 2	64	—	19	19	12	103
E/EA/CO 3	79	—	14	7	21.5	110

[a] Perkin–Elmer DSC-2. Samples cooled at 10 °C/min from above melting temperature, followed by heating at 10 °C/min.
[b] Temperature at last peak of melting endotherm.

monomer systems are not available, it is useful to compare E/VA/CO control 1 with E/EA/CO sample 2. Ethylene contents (66 versus 64 wt%) EA and VA contents and ΔH are fairly similar. Yet, the melting point is 44°C higher for the acrylate containing material. It appears reasonable to interpret this result as being due to a less uniform copolymer structure in the case of E/EA/CO.

VI. CONCLUSIONS

1. Ethylene–ethyl acrylate–carbon monoxide terpolymers (E/EA/CO) were prepared in high molecular weights via free radical initiated high-pressure batch polymerization techniques.

2. Proton magnetic resonance (PMR) was successfully utilized to determine composition of the terpolymer. Information concerning the sequence distribution of the terpolymers was also derived from the PMR spectra. A content of >4 wt% is necessary to observe a CO–ethylene–CO triad.

No evidence for a CO–EA–CO sequence was detected in any of the terpolymers.

3. Infrared spectra showed two carbonyl groups at 5.75 μm and 5.85 μm due, respectively, to the acrylate and carbon monoxide moieties.

4. Differential scanning calorimetry measurements indicate that E/EA/CO terpolymers retain a higher melting point than E/VA/CO terpolymers, probably due to the more random nature of the VA system.

APPENDIX. PMR CALCULATION OF TERPOLYMER CONTENT—
ETHYLENE–ETHYL ACRYLATE–CARBON MONOXIDE TERPOLYMERS

Assume the three monomeric units present are

$$+CH_2CH_2\rangle_m \quad +CH_2CH\rangle_n \quad +CH_2\overset{\overset{\displaystyle O}{\|}}{C}CH_2\rangle_o$$

$$\underset{\underset{\displaystyle OCH_2CH_3}{|}}{\overset{\displaystyle |}{C=O}}$$

Let:

A = area under peak at 4.17 ppm

B = area under peaks from 2.05–2.95 ppm

C = area under peaks from 0.5–2.05 ppm

The mole percent equations are:

$$\text{mol}\% \text{ EA} = (4A/(4B + 2C - 3A))(100)$$
$$\text{mol}\% \text{ CO} = ((2B - A)(4B + 2C - 3A))(100)$$
$$\text{mol}\% \text{ E} = (2(B + C - 3A)/(4B + 2C - 3A))(100)$$

REFERENCES

1. N. L. Zutty, J. A. Faucher, and S. Bonotto, *Encycl. Polym. Sci. Technol.* **6**, 387 (1967).
2. P. Ehrlich and G. A. Mortimer, *Adv. Polym. Sci.* **7**, 386 (1970).
3. N. L. Zutty and R. D. Buckhart, "Copolymerization" (G. E. Ham, ed.), Chapter XI. High Polymer Series **17**. Wiley (Interscience), New York, 1964.
4. F. P. Reding, J. A. Faucher, and R. D. Whitman, *J. Polym. Sci.* **57**, 483 (1962).
5. M. Matzner, D. L. Schober, R. N. Johnson, L. M. Robeson, and J. E. McGrath, "Permeability of Plastics, Films and Coatings" (H. B. Hopfenberg, ed.), p. 125. Plenum, New York, 1974.
6. R. A. V. Raff, *Encycl. Polym. Sci. And Technol.* **6**, 275 (1967).
7. N. L. Zutty and J. A. Faucher, *J. Polym. Sci.* **60**, 536 (1962).
8. G. Pieper, *Encycl. Polym. Sci. Technol.* **9**, 397 (1967).
9. J. E. Guillet, J. Dhanray, R. J. Golemba, and G. H. Hartley, *Adv. Chem. Ser.* **85**, 272 (1968).
10. J. E. Guillet, *Proc. IUPAC, Rio de Janiero, 1974* (E. Mano, ed.), p. 67. Elsevier, Amsterdam, 1975.
11. K. Tsuji, *Fortschr. Hochpolym.-Forsch.* **12**, 131 1973; *Chem. Abstr.* **80**, 15224b (1974).
12. S. Russo and S. Munari, *J. Polym. Sci., Part B* **5**, 827 (1967).
13. S. Russo, S. Munari, and E. Biagini, *J. Phys. Chem.* **73**, 378 (1969).
14. G. C. Alfonso, E. Fiorina, E. Martuscelli, E. Pedemonte, and S. Russo, *Polymer* **14**(8), 373 (1973).
15. C. W. Bunn, *J. Appl. Phys.* **25**, 820 (1954). [See also C. W. Bunn and H. S. Peiser, *Nature* **161**, 159 (1947).]
16. M. M. Brubaker, U. S. Patent 2,495,286 (to du Pont), 1950.

17. C. F. Hammer, French Patent 2,148,496 (to du Pont), 1973; *Chem. Abstr.* **79**, 93072u (1973).
18. C. F. Hammer, German Patent 2,238,555 (to du Pont), 1973.
19. L. M. Robeson and J. E. McGrath, paper presented at *Atlantic City AIChE Meet., 1976*; *Polym. Eng. Sci.* **17**(5), 300 (1977).
20. M. Matzner, L. M. Robeson, E. W. Wise and J. E. McGrath, *Amer. Chem. Soc., Div. Org. Coatings Plast. Chem. Prepr.* **37**(1), 123 (1977).

5

Carbon-13 Spin Relaxation
Parameters of Bulk Synthetic Polymers

RICHARD A. KOMOROSKI   *LEO MANDELKERN*

DIAMOND SHAMROCK CORPORATION   DEPARTMENT OF CHEMISTRY AND
PAINESVILLE, OHIO   INSTITUTE OF MOLECULAR BIOPHYSICS
                             FLORIDA STATE UNIVERSITY
                             TALLAHASSEE, FLORIDA

I. INTRODUCTION

The magnetic resonance parameters that are primarily related to molecular dynamics are the relaxation times T_1 and T_2. These quantities are the time constants governing the return of the z and x,y components respectively, of the macroscopic magnetization to their equilibrium values after a perturbation has been applied to the spin system [1]. The nonradiative processes of spin relaxation are governed primarily by mechanisms that consist of local magnetic (and sometimes electric) fields modulated by molecular motions (rotational and/or translational) of the proper frequency spectrum. Spin-lattice (T_1) processes are sensitive to motions at or near the nuclear larmor frequencies, which are typically 5–500 MHz. Spin–spin (T_2)

57

processes are sensitive to both low frequency motions (i.e., near zero) as well as to motions near the larmor frequency.

There are several possible mechanisms that can contribute to the spin relaxation of ^{13}C nuclei when they are present in natural abundance [2]. The predominant process for protonated carbons in polymers [3], and in all but the smallest molecules [4], is the ^{13}C–^{1}H dipolar interaction. Other mechanisms can be safely ignored. The ^{13}C dipolar relaxation is also the most useful mechanism to analyze molecular rotational motions. For reasons that will become evident shortly, the spin relaxation behavior of protonated carbons is not sensitive to translational motions. Moreover, sizable ^{13}C–^{13}C scalar interactions are so infrequent in the natural abundance experiment that they can be ignored.

The observation of ^{13}C spectra is routinely carried out under conditions of complete scalar proton decoupling. A number of advantages arise from this procedure. In addition to the collapse of each proton coupled multiplet into a single, narrow resonance, the saturation of the proton spins accompanying decoupling also results in a non-Boltzmann distribution of the ^{13}C spins in their energy levels [5]. This phenomenon, called the nuclear Overhauser enhancement (NOE), depends on the extent to which the ^{13}C–^{1}H dipolar interaction is the operative relaxation mechanism. It also depends on rotational motions near the larmor frequency, but in a manner different from that of T_1. For a carbon that is fully relaxed by the ^{13}C–^{1}H dipolar mechanism, up to a threefold signal enhancement can be obtained.

An important additional benefit accompanying proton saturation is the relatively simplified mathematical treatment that results. The recovery of the z and x,y magnetization components are found to be exponential, with first-order time constants T_1 and T_2 respectively [5]. It can be shown that the ^{13}C T_1, T_2, and nuclear Overhauser enhancement factor (NOEF = NOE − 1) are given by [2,4]

$$1/T_1 = 1/10\gamma_C^2\gamma_H^2\hbar^2 \sum_i r_i^{-6}[f(\omega_H - \omega_C) + 3f(\omega_C) + 6f(\omega_H + \omega_C)] \quad (1)$$

$$1/T_2 = 1/(2T_1) + 1/20\hbar^2\gamma_C^2\gamma_H^2 \sum_i r_i^{-6}[4f(0) + 6f(\omega_H)] \quad (2)$$

$$\text{NOEF} = (\gamma_H/\gamma_C)(6f(\omega_H + \omega_C) - f(\omega_H - \omega_C))$$
$$\div (f(\omega_H - \omega_C) + 3f(\omega_C) + 6f(\omega_H + \omega_C)) \quad (3)$$

Here γ_C and γ_H are the carbon and proton gyromagnetic ratios, respectively, ω_C and ω_H are the corresponding resonance frequencies, and r_i is the distance between the ^{13}C nucleus of interest and the ith relaxing proton. The spectral density functions $f(\omega_i)$ are Fourier transforms of the autocorrelation functions of second-order spherical harmonics. They describe the power available at angular frequency ω_i from the fluctuating interaction. The form of the $f(\omega_i)$ depends on the model used to describe the molecular motion. Assuming that only isotropic rotational diffusion of the CH vector(s) takes place, then $f(\omega_i)$ is given by the relation [2]

$$f(\omega) = \tau_R/(1 + \omega^2\tau_R^2) \tag{4}$$

where τ_R is the rotational correlation time. Here "isotropic" signifies that all spatial angles are equally available to the CH vector.

In Fig. 1 are plots of T_1, T_2, and NOEF versus τ_R for an isolated CH vector rotating isotropically in a field of 63.43 kG. There are several major features characteristic of the curves in Fig. 1. A given value of T_1 will correspond to two values of the correlation time (except at the minimum). These correspond to the so-called "fast" and "slow" solutions, depending on the value of τ_R. On the other hand, both T_2 and NOEF monotonically decrease with increasing τ_R. The correlation time region where $T_1 = T_2$ is called the region of extreme narrowing, and for the isotropic rotation of protonated carbons

$$1/T_1 = 1/T_2 = N\hbar^2\gamma_C^2\gamma_H^2 r_{CH}^{-6}\tau_R \tag{5}$$

and

$$\text{NOEF} = 2 \tag{6}$$

since $(\omega_H + \omega_C)\tau_R \ll 1$ in this region. Here N is the number of attached hydrogens and r_{CH} is the carbon–hydrogen bond distance. Implicit in Eq. (5)

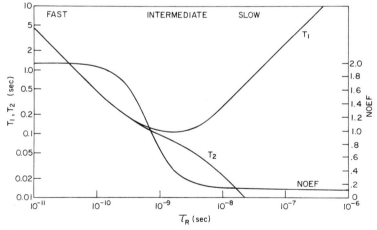

FIG. 1   Plots of T_1, T_2, and NOEF versus τ for an isolated CH vector rotating isotropically in a field of 63.43 kG. The bond distance r_{CH} is 1.09 Å. The T_1 and T_2 versus τ_R plots are log–log while the NOEF versus τ_R plot is semilogarithmic.

is the assumption that only directly bonded hydrogens are contributing to a given carbon relaxation behavior. In view of the r^{-6} dependence of the dipolar interaction, this assumption is a good one. Hence, for all motional considerations being equal, T_1 depends inversely on the number of attached hydrogens, while NOEF does not depend on this factor. With increasing resonance frequency, the range of correlation times that satisfy the extreme narrowing condition is reduced. In extreme narrowing, all relaxation parameters are independent of frequency. Outside of extreme narrowing, T_1 decreases with decreasing ω_C.

If molecular reorientation is anisotropic, with rigidly fixed CH vectors, or if the molecule undergoes internal motions, Eqs. (1)–(3) are still valid. However, now the spectral densities will depend on all components of the rotational diffusion tensor [6,7]. As a result, more than one correlation time is necessary to describe the complete motional behavior. Depending on the situation, the behavior of T_1 and NOEF can become quite complex [8].

Typical organic molecules ($M \gtrsim 500$) in nonviscous solvent usually having correlation times for overall rotation in the extreme narrowing region [2]. Most naturally occurring macromolecules such as the globular proteins as well as highly structured (i.e., hydrogen bonded) liquids at low temperatures quite often have correlation times in the "intermediate" and "slow" regions of the model curves given in Fig. 1. The above discussion has been concerned with the motions of the complete molecule. Although molecular reorientation could be a factor in dilute solution and perhaps for very low molecular weights in the pure state, it should not be a concern for high molecular weight chains in bulk. The predominant motional feature would be expected to be segmental mobility. For chain molecules, in order to observe the influence of segmental mobilities of the ^{13}C spin relaxation parameters the correlation times governing these motions must be comparable to or smaller than those characteristic of overall molecular reorientation. Such segmental motions must also occur at rates comparable to or smaller than the larmor frequency.

Previous spin relaxation results for several polymer systems in bulk and in moderately concentrated solution could not be interpreted on the basis of one or only a few correlation times [3,9–11]. These results are in accord with intuitive expectations, since for chain molecules segmental motions would be expected to encompass a broad range in correlation times. In the works cited, reduced NOEFs and relatively large line widths were observed while the T_1's apparently fulfilled the extreme narrowing condition for an isotropic rotation model. These apparently conflicting results, relative to small molecule behavior were interpreted by postulating a broad distribution of correlation times for isotropic rotation. One might expect that the motion of any given repeating unit would be modulated by many cooperative interactions between units. Thus, a given type of motion may involve the simultaneous cooperative action of one, several, or many segments. Presumably the correlation times describing such motion would increase with the increasing number of units involved. Details of the analysis, for a particular distribution function, are given elsewhere [3,10]. The T_1, T_2, and NOEF curves that result have the same general qualitative appearance as those shown in Fig. 1 for the single correlation time model. However, because of the distribution function the T_1 and NOEF curves flatten out as the breadth of the distribution is increased [3,10].

Experimentally, T_1's are readily measured for each resolved resonance from partially relaxed Fourier transform spectra [12]. Nuclear Overhauser enhancements are generally measured using a gated decoupling technique [13]. Values of T_2 are obtained from measured line widths $W_{1/2}$ since $T_2 = 1/(\pi W_{1/2})$.

II. TOTALLY AMORPHOUS POLYMERS

Before examining the structurally more complex semicrystalline polymers that have been studied it was deemed advisable to systematically investigate the spin-relaxation parameters of completely amorphous polymers as a function of molecular weight, temperature, chemical structure, and nmr frequency. Only a few isolated studies have been previously reported with no systematic variation, or control, of these primary variables.

A. Polyisobutylene and Butyl Rubber

Polyisobutylene was chosen for initial study since it is completely amorphous, is available over a very wide range in molecular weights, has a well-defined glass temperature that is well below ambient temperature, and is not subject to stereochemical complications. The complete results of this study have been reported elsewhere [14] and we limit ourselves here to a review of the major features.

The ^{13}C T_1's and NOEFs for each of the carbons of bulk polyisobutylene at 45°C and 67.9 MHz are given in Table I for a 3000-fold range in molecular weight. Because of the high molecular weights and the fact that a bulk system is being studied, the spin relaxation parameters will be the overwhelming result of backbone and side-chain segmental motions. Overall molecular motion will be too slow to be observed in this case. However, the effect of overall motion can be observed for relatively low molecular weight polymers ($M \gtrsim 2000$) in moderately concentrated or dilute solution [14,15].

As is indicated in Table I, both the ^{13}C T_1's and NOEFs are independent of molecular weight between 1.35×10^3 and 3.5×10^6. This molecular

TABLE I

Some ^{13}C Relaxation Parameters for Bulk Polyisobutylene at 67.9 MHz and 45°

	CH$_2$		C$_{quat}$		CH$_3$	
$MW \times 10^{-3}$	T_1 (sec)[a]	NOEF[c]	T_1 (sec)[b]	NOEF[c]	T_1 (sec)[b]	NOEF[d]
1.35	0.142	0.38	1.49	0.52	0.123	1.37
2.65	0.157	0.49	1.61	0.62	0.117	1.38
45.0	0.163	0.48	1.52	0.62	0.119	1.37
1000	0.176	0.40	1.57	0.86	0.123	1.24
3500	0.186	0.42	1.63	0.63	0.115	1.39

[a] Estimated accuracy, ± 10–15%.
[b] Estimated accuracy, $\pm 10\%$.
[c] Estimated accuracy, ± 0.2.
[d] Estimated accuracy, ± 0.1.

weight range encompasses eight orders of magnitude in the macroscopic viscosity. Therefore, the relatively fast and intermediate motions that determine these relaxation parameters are not influenced by segmental packing, the macroscopic viscosity, chain entanglements, and related phenomena. Hence, only very localized motions of the repeating units contribute to these relaxation parameters.

The T_1 values in Table I are typical of carbons of similar types in other polymers [3,9,11]. Those for the nonprotonated carbons are considerably longer than those for the protonated ones with the same motional behavior because of the sixth-power dependence of T_1 on the distance between the ^{13}C nucleus and the relaxing protons [4]. The quaternary carbon of polyisobutylene emphasizes this point.

The NOEFs of the backbone carbons of polyisobutylene are substantially smaller than seen for the bulk elastomers previously studied [3,9,10] at a lower frequency. In fact, they approach the theoretical minimum of 0.15, if only ^{13}C–H dipolar relaxation is operative. The NOEF of the side-chain methyl carbon, although definitely less than the allowed maximum value of 2, is substantially greater than that of the backbone carbons. This result can be attributed to the additional degree of freedom present for the methyl group, a fact not readily apparent for the corresponding T_1 values.

The resonance line widths are plotted against log molecular weight in Fig. 2 for the protonated carbons of polyisobutylene. Substantially reduced line widths are seen for all carbon for molecular weights of the order of several thousand. Above about $M = 5 \times 10^4$, the molecular weight does not effect the ^{13}C spectrum. Hence the slower modes of segmental motion (those motions involving more than just one or a few segments), characterized

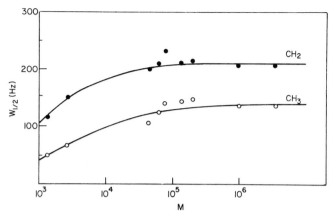

FIG. 2   Semilogarithmic plot of line width versus molecular weight for the protonated carbon resonances of bulk polyisobutylene at 45°C and 67.9 MHz.

by long correlation times and contributing to T_2 and hence the line width, do not change with molecular weight above a low critical value. Hence, except for the lower molecular weights, all of the relaxation parameters are independent of chain length for the completely amorphous polymer.

It might be expected that the introduction of intermolecular crosslinks would influence segmental motions and hence alter the ^{13}C spin relaxation parameters. However, for butyl rubber at typical crosslinking levels of about 0.2 crosslinks per 100 monomer units, the ^{13}C T_1's and the line widths are the same as for noncrosslinked polyisobutylene [16].

As was discussed previously, an isotropic model, i.e., one correlation time, for segmental reorientation does not adequately simultaneously explain the T_1, T_2, and NOEF data for the amorphous polymers previously studied [3,9]. To resolve the problem Schaefer introduced a broad, asymmetric distribution of correlation times for isotropic rotation [3,10]. A χ^2 distribution, characterized by a width parameter p, a mean correlation time $\bar{\tau}$, and a logarithmic time scale, was adopted for this purpose. In this formulation broad distributions are characterized by low values of p and the larger p the narrower the distribution function. The χ^2 distribution was chosen [10] since it allowed for a facile inclusion of the long correlation times necessary to describe cooperative motions among polymer segments. This distribution function, as well as others that have been subsequently introduced, are phenomenological in character, and were devised to explain the experimental data. However, they do not as yet have any molecular basis.

A similar analysis for polyisobutylene, at the operating frequency of 67.90 MHz, reveals that the data could not be explained using a "fast" solution for $\bar{\tau}$ with $p \geq 6$. However, with a "slow" solution for $\bar{\tau}$ (2×10^{-8} sec)

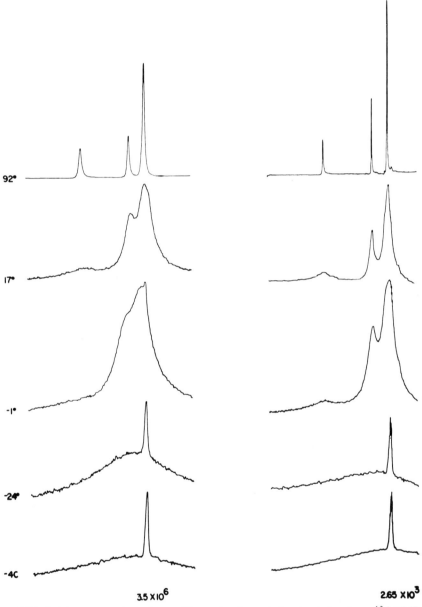

92°

17°

-1°

-24°

-4C

3.5 × 10^6 2.65 × 10^3

FIG. 3   Temperature dependence of the proton-decoupled, natural-abundance ^{13}C NMR spectra of bulk polyisobutylene at two molecular weights. A 5000-Hz region is shown. The spectra are plotted out at different vertical gain. The relatively narrow resonance apparent at low temperatures arises from the methyl carbon septet of acetone-d_6, which was used for external field-frequency locking.

and a p value of about 30, the T_1, T_2, and NOEF data could be satisfactorily accounted for. The p value of 30 indicates a relatively narrow distribution, which is not far removed from a one correlation time model [14]. The validity of the "slow" solution has been confirmed by studying the frequency dependence of T_1 [14].

The influence of temperature on the ^{13}C spectrum of bulk polyisobutylene at two extremes in molecular weight having the same glass temperature is illustrated in Fig. 3. Each of the carbon resonances broadens with decreasing temperature, merges with one another, and eventually cannot be observed. The low molecular weight sample yields a resolved spectrum at a temperature about 20° lower than is attainable for the high molecular weight one. No "high resolution" spectra are observed at temperatures less than about 70° above the glass temperature for all molecular weights. The temperature dependence of the spectrum of the low molecular weight is more pronounced than that of the high molecular weight.

At this point we should note that the glass temperature of *cis*-polyisoprene is the same as for polyisobutylene. However, for a high molecular weight sample of *cis*-polyisoprene resolvable spectra can be obtained 20–30° lower than polyisobutylene [17]. There is, therefore, no simple relation between the glass temperature and the temperature at which resolvable spectra can be obtained. However, at 40°C there are two orders of magnitude difference in backbone segmental mobility between polyisobutylene and *cis*-polyisoprene [14,17].

B. *cis*-Polyisoprene

A detailed study of the relaxation parameters for *cis*-polyisoprene in the completely amorphous state at 40°C and 22.6 MHz has been reported [9]. The data were interpreted in terms of the distribution of correlation times described previously. A good check can be obtained on the details of this analysis from experiments carried out at a substantially higher frequency. For this purpose and also to serve as a basis for the low temperature study of the supercooled and semicrystalline polymer, we have measured T_1, $W_{1/2}$, and NOEF at 40° and 67.9 MHz for *cis*-polyisoprene. The results are given in Table II and the data obtained at the lower frequency are also given for comparison.

All of the parameters, for each of the carbons, are different at the two frequencies. The larger T_1's and small NOEFs observed at 67.9 MHz are qualitatively expected from a one correlation time model [8,18]. Use of the distribution function and parameters employed by Schaefer at 22.6 MHz [10] yields values at 67.9 MHz that are given in parentheses and can be seen to be in excellent agreement with the experimental results. Since this

TABLE II

Carbon-13 Spin Relaxation Parameters for cis-Polyisoprene at 40°C

Carbon[f]	22.6 MHz[a]			67.9 MHz[e]		
	T_1^b	NOEF[c]	$W_{1/2}^d$	T_1^b	NOEF[c]	$W_{1/2}^d$
α	700	1.2	7	940	0.4	38
β	95	1.2	14	166(200)	0.7(1.0)	41
γ	50	1.2	20	114(100)	0.9(1.0)	41
δ	55	1.2	18	119(110)	0.8(1.0)	40
ε	350	1.2	12	656	0.9	36

[a] Reprinted with permission from R. A. Komoroski, J. Maxfield, and L. Mandelkern, *Macromolecules* **10**, 545 (1977). Copyright by the American Chemical Society.

[b] In msec. Estimated accuracy $\pm 10\%$.

[c] Estimated accuracy ± 0.1.

[d] In Hz. Estimated accuracy $\pm 10-15\%$.

[e] Values in parentheses are predicted from data at 22.6 MHz and a distribution of correlation times model (see text).

[f] The designation of the carbon atoms is $-CH_2-C(CH_3)=CHCH_2-$ $\gamma,\alpha,\varepsilon,\beta,\delta$.

agreement is for a specific polymer at a given temperature, caution needs to be exercised in accepting the general validity of this particular distribution function.

The line widths at 67.9 MHz, however, are predicted to be about 7% lower than at 22.6 MHz. Table II reveals that this is clearly not the case. The discrepancy in line width could result from inadequacies in the distribution function or from nonmotional contributions such as field inhomogeneity or macroscopic bulk susceptibility differences in the irregularly configured sample. The contribution from the latter factors depends linearly on field strength and would be six times more pronounced at the higher field being used here [19,20].

III. CHEMICAL STRUCTURE AND SEGMENTAL MOTION

Table III gives a comparison of the backbone motions of several completely amorphous polymers. The three polymers chosen have very similar glass temperatures, and are elastomers at the temperature of measurements, 40–45°, which is at least 100° above the glass temperature. Despite these similarities the segmental mobilities of the backbone carbons are quite different. The average correlation time for segmental motion varies by more

TABLE III

Comparison of Backbone Motions of
Several Elastomers at 40–45°C

	T_g(°C)	$\bar{\tau} \times 10^9$ sec	p
Polyisobutylene[b]	− 70	20	30
cis-Polyisoprene[a]	− 70	0.4	14
cis-Polybutadiene[a]	− 85	0.01	9

[a] For methine carbon.
[b] For methylene carbon.

than three orders of magnitude. In addition, in terms of the Schaefer distribution function, as the average mobility increases a broader distribution of correlation times is required to explain the data. These results make clear that the ^{13}C spin relaxation parameters of amorphous polymers, well above T_g, are not related to any simple macroscopic property. They do point out that the chemical nature of the repeating unit has a major influence on these quantities. Clearly, these studies are in their most formative stages. More data for widely different chemical types, and as a function of temperature, are needed before any generalizations can be made and the relationship to other properties established.

IV. SEMICRYSTALLINE POLYMERS

The molecular structure of a semicrystalline polymer is very complex [21,22]. A lamella-like crystallite is the usual morphological feature of homopolymers. Associated with these crystallites are interfacial as well as interzonal or amorphous regions. The crystallites can be organized further, under certain circumstances, into higher levels of supermolecular structure [22]. Contrary to widespread feeling, a spherulitic morphology is not universally observed, but the superstructure depends on molecular weight, polydispersity, and crystallization temperature [22]. Since ^{13}C spin relaxation parameter measurements can provide detailed information about segmental motions, it is appropriate to apply these techniques to semicrystalline polymers to explore the influence of these various structural parameters. It becomes mandatory that in order to obtain meaningful results the different structural factors have to be systematically controlled. It should be pointed out in this connection that the measurements described here probe the motions in only the amorphous and/or interfacial regions of the polymer, i.e., the noncrystalline regions. We chose for this initial study cis-polyisoprene and linear polyethylene.

A. *cis*-Polyisoprene

For semicrystalline polymers there is a fundamental question as to whether there are any differences between the structure and properties of the noncrystalline regions of the polymer and the completely liquid (amorphous) state of the same polymer at the same temperature and pressure [21]. For the present type of experiments such a comparison is possible for *cis*-polyisoprene, which is a sufficiently slowly crystallizing polymer so that it can be studied, in independent experiments, in both the semicrystalline and totally amorphous states at 0 to $-10°C$.

Table IV compares at $0°C$ the ^{13}C T_1's and NOEFs for all carbons of *cis*-polyisoprene in a completely amorphous sample with the same polymer having a degree of crystallinity 0.31. Table V contains the resonance line widths for the completely amorphous samples at $0°C$ and $-10°C$ and two semicrystalline samples $(1 - \lambda = 0.12$ and 0.31). Figure 4 illustrates the ^{13}C spectra at $0°$ as a function of the level of crystallinity.

For the completely amorphous sample, at $0°$ the T_1's of all the carbons are longer than the comparable value at $40°$, suggesting that the "slow" solution [8] for $\bar{\tau}$ is applicable at $0°$. Except for the methyl carbon, all the NOEFs at $0°$ are reduced relative to their value at $40°$. These results are consistent with the expected reduction in backbone mobility at the lower

TABLE IV

*Carbon-13 Spin-Lattice Relaxation Times and
Nuclear Overhauser Enhancement Factors for
cis-Polyisoprene at 0°C and 67.9 MHz[a]*

Carbon	Amorphous		Semicrystalline[f]	
	T_1 (msec)[b]	NOEF[d]	T_1 (msec)[b]	NOEF[d]
α	2000	0.1	2,400	0.1
β	400	0.3[e]	460	0.3[e]
γ	260[c]	0.3	270[c]	0.4
δ	240[c]	0.3	270[c]	0.4
ε	840[c]	0.8	910[c]	1.2

[a] Reprinted with permission from R. A. Komoroski, J. Maxfield, and L. Mandelkern, *Macromolecules* 10, 545 (1977). Copyright by the American Chemical Society.

[b] Estimated accuracy $\pm 10\%$, except where noted.

[c] Estimated accuracy $\pm 15\%$.

[d] Estimated accuracy ± 0.2, except where noted.

[e] Estimated accuracy ± 0.1.

[f] Degree of crystallinity equals 0.31.

TABLE V

Carbon-13 Line Widths[a] (Hz) for cis-Polyisoprene
at 0 and $-10°C$ and 67.9 MHz[b]

Carbon	Amorphous		Semicrystalline (1): $1 - \lambda = 0.12$	Semicrystalline (2): $1 - \lambda = 0.31$	
	0°C	$-10°C$	0°C	0°C	$-10°C$
α	70	90	92	108	290
β	114	205	142	178	500
γ	165[c]		197[c]	248[c]	
δ	124[c]		146[c]	170[c]	
ε	69[d]	85	96[d]	120[d]	250

[a] Estimated accuracy ± 4 Hz, except where noted.

[b] Reprinted with permission from R. A. Komoroski, J. Maxfield, and L. Mandelkern, Macromolecules 10, 545 (1977). Copyright by the American Chemical Society.

[c] Estimated accuracy $\pm 30\%$

[d] Estimated accuracy $\pm 10\%$

temperature. A distribution of correlation times analysis for the data of the totally amorphous polymer at 0° yields values of ~ 30 and 3.2×10^{-8} sec for p and $\bar{\tau}$, respectively. As mentioned above, a value of 14 for p was used at 40°. This result appears surprising, since it would be expected that the increased freedom introduced into the polymer backbone with increasing temperature would result in a narrower distribution. Apparently the shape of the distribution changes with temperature, i.e., different correlation time regions possess different activation energies.

From the point of view of crystalline samples it is found that both T_1 and NOEF are the same in the completely amorphous and semicrystalline sample at 0°. Thus the level of crystallinity achieved does not affect the relatively fast motions in the amorphous regions that are reflected in the T_1's and NOEFs. For polyethylene (see below), T_1 is independent of crystallinity between 0.51 and 0.94 [23]. However, the NOEF of polyethylene did decrease at the highest crystallinity levels from the maximum value of 2 observed at $1 - \lambda = 0.51$. Considering the results for both polymers, we can conclude that T_1 is invariant over the complete crystallinity range. The NOEFs observed for the backbone carbons of cis-polyisoprene at 0°C in the completely amorphous state are near the theoretical lower limit. They could, therefore, only be affected slightly by any changes in the width of the distribution function caused by crystallinity. Hence, they would be insensitive to changes in the level of crystallinity.

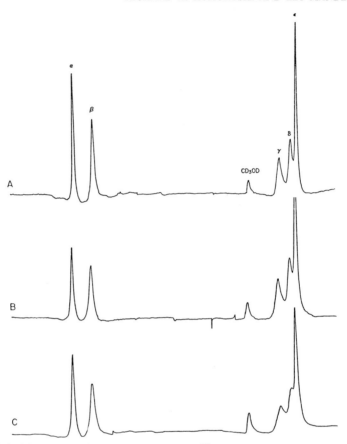

FIG. 4   Natural abundance proton-decoupled ^{13}C NMR spectra of amorphous and semicrystalline *cis*-polyisoprene at 0°C. About 5-Hz broadening, due to exponential filtering, is present in each case. An 11-kHz region is shown. (A) $1 - \lambda = 0.00$, 50 scans, 9.0 sec pulse delay; (B) $1 - \lambda = 0.12$, 256 scans, 5.0 sec pulse delay; (C) $1 - \lambda = 0.31$, 250 scans, 9.0 sec pulse delay. (Reprinted with permission from R. A. Komoroski, J. Maxfield, and L. Mandelkern, *Macromolecules* **10**, 545 (1977). Copyright by the American Chemical Society.)

However, in contrast to the T_1's and NOEFs, the line widths depend on the level of crystallinity, as is shown in Table V and Fig. 4. As the level of crystallinity increases there is a significant monotonic increase in line width, which is observed even at the lowest level of crystallinity studied. Line broadening of the same order of magnitude has been reported for non-crystalline *cis*-polyisoprene filled with carbon black relative to the unfilled polymer [9]. The addition of carbon black filler also had little effect on T_1 and NOEF but shortened the methine carbon T_2 by a factor of about 5–10. The same form of the distribution function quantitatively accounted for these parameters. The value of $\bar{\tau}$ was the same as for the unfilled polymer, but the value of p was lowered from 14 to 10.

A formal analysis, similar to that for the filled polymer [10], can also be made for semicrystalline cis-polyisoprene $(1 - \lambda = 0.31)$ at 0°C. It is not possible simultaneously to explain the T_1, NOEF, and T_2 of the semicrystalline polymer by increasing $\bar{\tau}$ with p remaining constant. With $\bar{\tau}$ kept constant, and the distribution broadened, if p is reduced from 30 to about 20 then the calculated line width is 245 Hz, as compared with the observed 198 Hz. Obviously, changes in the parameters in the log χ^2 distribution can only serve as a guide to the segmental motion at the molecular level. This function should only be applicable when the long correlation times in the tail of the distribution are not in, or approaching, the rigid lattice region. These conditions need not necessarily be fulfilled for semicrystalline polymers, where longer correlation times in the noncrystalline region can be anticipated.

The results for semicrystalline cis-polyisoprene can be interpreted in a manner similar to that given by Schaefer for the filled system [10]. In the present case crystallinity could retard complete segmental motional narrowing. The residual dipolar interactions are associated with the longer correlation times. Crystallinity has little effect on either the T_1's or NOEFs of cis-polyisoprene, indicating that the relatively high frequency motions are not changed. These are clearly rather important results concerning segmental motions in semicrystalline polymers. It is thus important to ascertain the generality of these results, the influence on them of morphology, and their correlation with other properties. As a step in this direction we have also studied the relaxation properties of semicrystalline linear polyethylene [23], crystallized in the bulk, which can be prepared over a wide range in level of crystallinity [21,22] and in several distinctly different morphological forms [24].

B. Linear Polyethylene

The results of the linear polyethylene study are summarized in Table VI [23]. The corresponding morphological forms were obtained from density and the viscosity average molecular weight of the sample. As is indicated, in many cases molecular weight fractions were used, and the relaxation parameters that were determined pertain to the noncrystalline regions.

We can now examine, in turn, the dependence of the relaxation parameters on the quantities characterizing the crystallinity and molecular weight, as well as on the different morphological forms and spherulite sizes [23]. T_1 is found to be constant and equal to about 350 msec. The invariance of T_1, previously observed for cis-polyisoprene, can now be considered to be valid

TABLE VI

Carbon-13 Spin Relaxation Parameters of Linear Polyethylene at 45°C and 67.9 MHz

M_η	$(1 - \lambda)^g$	Morphology	T_1 (msec)^a	Line Width $W_{1/2}$ (Hz)^b	NOEF
8.1×10^4	.57	Spherical	343	622	—
2.5×10^5	.51	Spherical	355	625	2.0^c
1.7×10^{5f}	.81	Spherical	348	695	1.5^d
1.7×10^{5f}	.68	Spherical	356	700	—
2.0×10^{6f}	.51	None	369	501	2.0^c
2.0×10^{6f}	.72	None	358	496	—
6.1×10^6	.54	None	—	500^e	—
6.1×10^6	.70	None	—	503	—
$2.7_5 \times 10^4$	.94	Rod	352	945	1.0^d

^{a} Estimated accuracy $\pm 10\%$.
^{b} Estimated accuracy $\pm 5-10\%$.
^{c} Estimated accuracy ± 0.1.
^{d} Estimated accuracy ± 0.2.
^{e} Estimated.
^{f} Unfractionated sample.
^{g} Obtained from density measurements.

over the complete crystallinity range. Thus the relatively fast, and hence localized, segmental motions that determine T_1 are the same for all non-crystalline regions of the samples studied. This result could be intuitively anticipated for samples of low density that do not form spherulites, since long sequences of disordered units are present. The fact that the spherulitic samples also have the same T_1 indicates that the fast segmental motions of the amorphous material within the system of organized lamellae are also the same. In addition, it is quite surprising that the 94% crystalline sample, of very high density, has the same T_1. In this case the signal must arise almost entirely from regions that are on or near the crystallite surface, and might be anticipated to be highly constrained. The T_1 value and full NOEFs for the samples of lowest density suggest a "fast" solution for the correlation time for segmental motion.

It is informative to compare the segmental motions of the noncrystalline regions of linear polyethylene, as monitored by ^{13}C T_1's, with those of molten, low molecular weight n-alkanes. The T_1 for polyethylene is one-third to one-half the value found for the interior carbons of molten n-alkanes of 18–20 carbon atoms [25,26]. Lyerla et al. [26] used a self-consistent analysis to determine correlation times for internal motion. The effective correlation time, τ_{eff}, is given by Eq. (5) (in place of τ_R). The value of τ_{eff} for the jth carbon within the chain was considered to have contributions from both overall an internal motions

$$(^j\tau_{\text{eff}})^{-1} = (^j\tau_i)^{-1} + (\tau_o)^{-1} \tag{7}$$

Here τ_o is the isotropic overall correlation time (the same for all carbons) and $^j\tau_i$ is the correlation time for internal motion of the jth carbon. The difference in the rates of segmental motion between carbons j and k is then

$$\tau(j,k)^{-1} = (^j\tau_{\text{eff}})^{-1} - (^k\tau_{\text{eff}})^{-1} = (^j\tau_i)^{-1} - (^k\tau_i)^{-1} \tag{8}$$

For the largest n-alkane studied to date, n-eicosane ($C_{20}H_{42}$) [26], up to four carbons from the chain end could be resolved. Using this treatment, the best available estimate of backbone internal motion would be $\tau(4, \text{internal})$, which was found to be 6.6×10^{-11} sec for n-eicosane at 39°C. For polyethylene, if one identifies τ_{eff} with τ_{seg}, then from Eq. (5) and a T_1 of 350 msec we obtain the same value of the correlation time at 45°. While the exact quantitative agreement obtained between $\tau(4, \text{internal})$ of n-eicosane and τ_{seg} of polyethylene is probably fortuitous, and the use of an isotropic model for polyethylene is admittedly approximate, these calculations strongly suggest that the segmental motions are essentially identical in the two cases.

In contrast to the T_1's, the NOEFs given in Table VI are not the same for all samples. The two samples of lowest crystallinity (0.51) have the theoretically maximum NOEF of 2.0, as was also determined for molten polyethylene at 155° by Inoue [27]. For samples with the highest crystallinity, 0.81 and 0.94, the NOEF is significantly reduced to 1.5 and 1.0, respectively. Hence, these observations suggest that the NOEF depends on the level of crystallinity and not specifically on the morphological form. We note that in this case we are in a correlation time region where the NOE is more sensitive to changes than *cis*-polyisoprene.

In order to explain only the T_1 and NOEF results by themselves for the lowest crystallinity samples, it is not necessary to postulate a distribution of correlation times since full NOEFs are observed. The reduced NOEFs observed for the higher crystallinity samples could arise from some distribution phenomenon imposed by the crystalline region. The detailed nature of such a distribution is not clear at present. Evidently the noncrystalline regions of these samples experience chain motions that reduce the NOEF substantially but do not affect T_1 or T_2 to an extent expected on the basis of previous studies [3,9,10].

The experimentally observed ^{13}C line widths are not the same for all the polyethylene samples. For the experimental conditions and samples used here they are in the range of 500–900 Hz. For other polymers studied so far, both in bulk and in moderately concentrated solution, T_1 is not equal to T_2 [3,9,11]. The polyethylene line widths are about two to three orders of magnitude broader than would be expected if $T_1 = T_2$. Results such as these have been attributed in a very general way to the fact that for polymers the segmental motions contributing to T_1 and NOEF are not the same as those that determine T_2, and hence $W_{1/2}$. The slower modes of polymer motion do not effectively contribute to T_1 and NOEF, but can contribute to T_2. However, for elastomers with effective correlation times comparable to that of polyethylene, greatly reduced NOEFs and line widths on the order of 20–50 Hz were observed [9,10]. These are very narrow compared to those found for polyethylene.

A major factor contributing to the broad polyethylene line widths is the crystalline character. As was previously discussed for *cis*-polyisoprene, the noncrystalline regions of semicrystalline samples yield broader resonances than for the totally amorphous polymer. A study of *trans*-polyisoprene, which is about 40% crystalline at 40°, indicated a broadening from this cause also [9]. However, a direct comparison with the totally amorphous polymer to firmly establish this point for this polymer was not made. There is, therefore, a substantial body of evidence that demonstrates that crystallinity causes broader spectral lines. However, the polyethylene line widths are substantially larger than have been observed for other semicrystalline polymers at comparable temperatures. This is one of the major distinguishing features of this semicrystalline polymer. There is, therefore, a major problem, namely, to account simultaneously for the T_1's and full or substantial NOEFs, which suggest fulfillment or near-fulfillment of the extreme narrowing condition, while the line widths are an order of magnitude larger than observed for *trans*-polyisoprene [9] and thus would not be consistent with the above condition.

Several possible explanations can be offered for the abnormally large line widths. One is that the ^{13}C spin relaxation behavior arises from a distribution of correlation times (for isotropic rotation) phenomena. In view of the T_1's and full or substantial NOEFs, a distribution resembling that used previously by Schaefer [10] for amorphous elastomers can not simultaneously explain the data. In fact, such a distribution can not adequately explain the data for crystalline *trans*-polyisoprene [9,17]. Phenomenologically, it should be possible to construct quite easily a modified distribution that would simultaneously explain all three spin relaxation parameters. A distribution of bimodal form would probably suffice. However, a purely mathematical construction by itself would have little connection to any molecular interpretation.

Another possibility is that the large line widths are due to incomplete motional narrowing, arising from the presence of crystallites and of chain entanglements in the amorphous region. This possibility is to be distinguished from broadening due to slow, but isotropic, rotation. If the explanation of incomplete motional narrowing is accepted, it is still necessary to explain the major difference observed between *trans*-polyisoprene and at least the polyethylene of lowest crystallinity. These polymers have approximately the same degree of crystallinity, and on the basis of their T_1's both exhibit fast segmental mobility. If this explanation is acceptable then there must be a major influence of the chemical nature of the repeating unit.

A third possibility is that nonmotional or static phenomena are making a substantial contribution to the line widths. Differences in bulk magnetic susceptibility within the sample volume can result in differences in nuclear screening among nuclei in different regions of the sample, resulting in a broadening of the resonance [19,20]. Such broadening could occur from the irregular macroscopic sample configuration used here, or possibly from microscopic differences within the sample. Broadening from this cause will vary linearly with the applied field [19], and hence will be approximately six times more severe here at 63.4 kG than in all previous studies of this type, which were carried out at a much lower field strength [9].

A number of experimental procedures exist to determine whether or not bulk susceptibility broadening contributes measurably to an observed line width. In addition to the aforementioned frequency dependence experiment [19] one can perform so-called "hole burning" experiments [9,28] to ascertain whether a substantial portion of the line width results from inhomogeneous broadening. Experiments such as have been described above are currently in progress [29] in connection with the extraordinary broad resonance lines associated with the semicrystalline polyethylene.

Further insight into the basis for the broadening can be obtained by studying the temperature dependence of the relaxation parameters. It has been found that T_1 smoothly increases from 350 to 1590 msec between 45 and 145°. As expected, the line width decreases with increasing temperature. However, the value about 180 Hz for the line width at 145°, which is in the pure melt, is still substantially larger than expected based on the results for other amorphous polymers. The ratio of T_1 to the apparent T_2, as calculated from the measured line width, actually increases with increasing temperature. The usual expectation is that with increasing temperature T_1/T_2 will decrease and approach unity [30]. The behavior of T_1/T_2 for polyethylene is difficult to explain if only motional factors are responsible for the line width at 145°. At this temperature all restraint on chain mobility arising from the crystalline regions are absent. The influence of factors responsible for incomplete motional narrowing at 145° should be greatly reduced. Even though chain entanglements will still be present, the motional behavior of a CH vector in polyethylene at 145° should approach that in a low molecular weight *n*-alkane in the liquid state. It is possible that the shape of the correlation time distribution changes in such a way as to produce the observed T_1/T_2 behavior. This implies that different correlation time regions of the distribution have different activation energies. However, the activation energies of the longer (T_2) correlation times must be less than those of the short (T_1) correlation times. This last conclusion is intuitively unsatisfying. However, the presence of a large relatively static contribution to the apparent T_2 is consistent with the observed behavior.

The line width data in Table VI does not show any obvious correlation with the level of crystallinity or molecular weight polydispersity. However, based on the raw data there does appear to be a correlation with the morphology or supermolecular structure. The line widths fall into essentially three categories: those of about 500 Hz, which have no distinct morphology and are randomly oriented crystallites; those with definite spherulitic morphology whose line widths range from about 600–700 Hz and may also depend on the spherulitic radius; and those with about a 950-Hz line width, which are associated with the rod-like morphology of the sample of highest density.

Detailed correlations between line widths of ^{13}C resonances arising from the noncrystalline regions of semicrystalline polymers and the morphology cannot be made with certainty, however, until the major factors contributing to the unusually broad lines associated with polyethylene have been established. If these initial observations are substantiated then an important influence of the supermolecular structure on the mechanical properties, transitions, and glass formation can be anticipated.

From these first studies we can conclude that the ^{13}C spin-relaxation parameters have the potential of providing important information concerning the influence of the chemical nature of the repeating unit, of the structure, and of the morphology on the properties of both completely amorphous and semicrystalline polymers. As has been indicated in the text, certain kinds of additional data and experiments are required as well as appropriate models relating segmental motion to molecular structure.

ACKNOWLEDGMENT

This work was supported by the National Science Foundation under Grant Number DMR76-21925.

REFERENCES

1. A. Abragam, "The Principles of Nuclear Magnetism." Oxford Univ. Press, London and New York, 1961.
2. J. R. Lyerla, Jr., and G. C. Levy, "Topics in Carbon-13 NMR Spectroscopy" (G. C. Levy, ed.), Vol. 1, Chapter 3. Wiley (Interscience), New York, 1974.

3. J. Schaefer, "Topics in Carbon-13 NMR Spectroscopy" (G. C. Levy, ed.), Vol. 1, Chapter 4. Wiley (Interscience), New York, 1974.
4. A. Allerhand, D. Doddrell, and R. Komoroski, *J. Chem. Phys.* **55**, 189 (1971).
5. K. F. Kuhlmann, D. M. Grant, and R. K. Harris, Jr., *J. Chem. Phys.* **52**, 3439 (1970).
6. W. F. Huntress, *Advan. Mag. Res.* **4**, 1 (1970).
7. D. E. Woessner, *J. Chem. Phys.* **36**, 1 (1962).
8. D. Doddrell, V. Glushko, and A. Allerhand, *J. Chem. Phys.* **56**, 3683 (1972).
9. J. Schaefer, *Macromolecules* **5**, 427 (1972).
10. J. Schaefer, *Macromolecules* **6**, 882 (1973).
11. J. Schaefer and D. F. S. Natusch, *Macromolecules* **5**, 416 (1972).
12. R. L. Vold, J. S. Waugh, M. P. Klein, and D. E. Phelps, *J. Chem. Phys.* **48**, 3831 (1968).
13. R. Freeman, H. D. W. Hill, and R. Kaptein, Jr., *J. Magn. Res.* **7**, 327 (1972).
14. R. A. Komoroski and L. Mandelkern, *J. Polym. Sci., Part C* **54**, 201 (1976).
15. A. Allerhand and R. K. Hailstone, *J. Chem. Phys.* **56**, 3718 (1972).
16. R. A. Komoroski and L. Mandelkern, *J. Polym. Sci., Part B* **14**, 253 (1976).
17. R. A. Komoroski, J. Maxfield, and L. Mandelkern, *Macromolecules* **10**, 545 (1977).
18. R. A. Komoroski, I. R. Peat, and G. C. Levy. *Biochem. Biophys. Res. Commun.* **65**, 272 (1975).
19. J. K. Becconsall, P. A. Curnuck, and M. C. McIvor, *Appl. Spectrosc. Rev.* **4**, 307 (1971).
20. J. A. Pople, W. G. Schneider, and A. Bernstein, "High-Resolution Nuclear Magnetic Resonance," p. 80. McGraw–Hill, New York, 1959.
21. L. Mandelkern, "Characterization of Materials in Research, Ceramics, and Polymers," Chapter 3. Syracuse Univ. Press, Syracuse, New York, 1975.
22. L. Mandelkern, *Accounts Chem. Res.* **9**, 81 (1976).
23. R. A. Komoroski, J. Maxfield, F. Sakaguchi, and L. Mandelkern, *Macromolecules* **10**, 550 (1977).
24. S. Go, R. Prud'homme, R. Stein, and L. Mandelkern, *J. Polym. Sci., Part A-2* **12**, 1185 (1974).
25. W. J. M. Birdsall, A. G. Lee, Y. K. Levine, J. C. Metcalfe, P. Partington, and G. C. K. Roberts, *J. Chem. Soc., D* 757 (1973).
26. J. R. Lyerla, Jr., H. M. McIntyre, and D. A. Torchia, *Macromolecules* **7**, 11 (1974).
27. Y. Inoue, A. Nishioka, and R. Chujo, *Makromol. Chem.* **168**, 163 (1973).
28. N. Bloembergen, E. M. Purcell, and R. V. Pound, *Phys. Rev.* **73**, 679 (1948).
29. D. Axelson, R. A. Komoroski, and L. Mandelkern, unpublished observation.
30. D. W. McCall, *Accounts Chem. Res.* **4**, 223 (1971).

6

NMR and Infrared Study of Thermal Oxidation of cis-1,4-Polybutadiene

ROBERT V. GEMMER* and MORTON A. GOLUB

AMES RESEARCH CENTER
NATIONAL AERONAUTICS AND SPACE ADMINSTRATION
MOFFETT FIELD, CALIFORNIA

I. INTRODUCTION

The oxidation of 1,4-polyisoprene has been studied extensively, while the corresponding process in 1,4-polybutadiene has received much less attention [1–4]. In recent years there have been attempts to redress this deficiency with mechanistic papers on the thermal [5], photochemical [6], and photosensitized [7] oxidations of cis-1,4-polybutadiene (CB), using infrared spectroscopy as the major analytical tool. Having previously reported the first nuclear magnetic resonance (NMR) spectroscopic study of the thermal oxidation of 1,4-polyisoprene [8], we deemed it desirable to follow up that work with a similar study on CB, with particular emphasis on using ^{13}C NMR

---

* National Research Council–National Aeronautics and Space Administration Research Associate 1975–1977. Present address: American Cyanamid Company, Stamford, Connecticut.

79

spectroscopy. The latter has a special advantage over ^{1}H NMR spectroscopy in being able to provide high resolution spectra for elastomers in the solid state [9]. This feature of ^{13}C NMR spectroscopy is thus of considerable value in the case of CB since this polymer is well known to crosslink on oxidation, with attendant insolubilization [1,4].

The major functional group to appear in the ^{13}C spectrum of oxidized CB was found to be epoxides. This finding contrasts with the situation in the thermal oxidation of 1,4-polyisoprene [8], where the formation of significant amounts of peroxide and alcohol groups in addition to epoxides was observed. The presence of alcohols and, in lesser amounts, of peroxides in oxidized CB was adduced from complementary infrared and NMR data on soluble extracts of the oxidized elastomer.

II. EXPERIMENTAL TECHNIQUE

A. Oxidations

The CB used in this work was obtained from The B. F. Goodrich Research and Development Center, Brecksville, Ohio. The preparation of solutions of CB and the thermal oxidation of CB films were similar to procedures described previously for 1,4-polyisoprene [8]. The film oxidations were run at temperatures from 90–180°C. The development with time of spectroscopic features associated with oxidation followed sigmoid curves characteristic of autoxidation [1–4]. Following oxidation a portion of the reaction product was used directly to obtain ^{13}C NMR spectra and elemental analyses for the total oxidized polymer, while the remainder was extracted with $CHCl_3$ under nitrogen using a Soxhlet apparatus. NMR and infrared (ir) spectra, as well as elemental analyses, were obtained from the extracts.

B. Spectroscopy

^{1}H and ^{13}C NMR spectra were obtained at ambient temperature on a Varian HA-100D or Varian CFT-20 NMR spectrometer as described previously [8]. Some ^{13}C NMR spectra were run at higher than ambient temperature using 8-mm tubes and perdeuteriobromobezene as solvent. Chemical shifts are reported in ppm relative to tetramethylsilane. Infrared spectra were obtained on a Perkin–Elmer 180 spectrophotometer from thin films deposited on NaCl plates.

C. Infrared Calibration

The absorbance of the alcohol peak at 2.9 μm relative to that of the 6.9-μm peak as an internal reference was measured for several hydroxy-containing CB samples of known alcohol content obtained by the reduction with NaBH$_4$ of the corresponding hydroperoxidized CB samples, previously prepared through singlet oxygenation [10]. It was thus determined that the absorbance ratio, $A'_{OH} \equiv A_{2.9}/A_{6.9}$, was proportional to alcohol content and had a value of 1.90 corresponding to 0.5 OH/monomer unit in the hydroxy-containing CB.

III. RESULTS

The ^{13}C NMR spectrum obtained for a typical sample of oxidized CB is presented in Fig. 1. Although run as a gel in C$_6$D$_6$, the spectrum is well resolved. Apart from the resonances at (A) 27.5 and (B) 129.4 ppm associated with unreacted monomer units [9] and the three resonances at 126–129 ppm due to the solvent, the only prominent peaks to appear in the region down-field from 40 ppm are assigned to epoxides.[†] This result was unexpected in view of the complex mixture of epoxides, peroxides and alcohols found in the ^{13}C NMR spectrum of oxidized 1,4-polyisoprene [8]. It was first thought that the failure to detect a significant amount of peroxides was due to local immobilization of the polymer chains caused by peroxide crosslinks that might serve to broaden the signals of carbons near the crosslinks while having little effect on the signals of carbons remote from them. Attempts to confirm this view by searching for a temperature dependence of the peroxide region (78–88 ppm) proved unsuccessful, with no increase in this region being detected up to 100°C. It was later found that the sol portion obtained by CHCl$_3$ extraction of the gel gave a ^{13}C NMR spectrum that showed the same features as the original gel. As the oxygen contents of the sol and gel were substantially the same, these observations effectively ruled out interference in the NMR due to locally immobilized chain segments. A possible alternative explanation for the lack of signals due to peroxides was that the nuclear Overhauser effect or relaxation time values for these carbons might have been abnormal. However, this possibility was dismissed in light of the normal behavior of similar peroxide-containing systems [8,10].

The assignment of the resonances at 56 and 58 ppm in Fig. 1 to cis and trans epoxides, respectively, follows from inspection of Fig. 2, which shows the ^{13}C NMR spectra of CB and *trans*-1,4-polybutadiene after partial epox-idation. The details of the syntheses and ^{13}C NMR spectroscopy of epox-idized polydienes will be reported elsewhere [11]. Besides peaks 4 and 4', due to the carbons bearing oxygen, other epoxide-related peaks expected to appear in the spectrum of oxidized CB are peaks 1, 3, and 3' of Fig. 2, but not peak 1' inasmuch as the cis and trans epoxides appear in an otherwise all-cis polymer. Thus, the peaks at 25 and 33 ppm in Fig. 1 correspond to peaks 1 and 3' of Fig. 2, respectively. Peak 3 of Fig. 2 is buried under peak A of

---

† See Fig. 2 and [8].

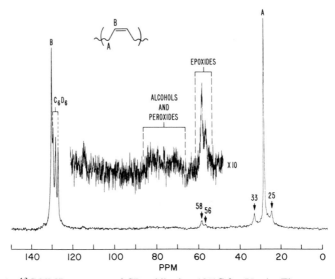

FIG. 1 ^{13}C NMR spectrum of CB oxidized at 125°C for 75 min. The spectrum was run as a gel in C_6D_6.

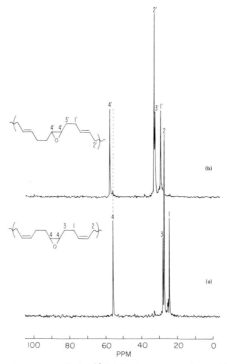

FIG. 2   The aliphatic region of the ^{13}C NMR spectra of epoxidation products from (a) *cis*- and (b) *trans*-1,4-polybutadiene.

82

Fig. 1, but was observed in the better resolved spectrum of the sol obtained from the gel used for Fig. 1.

The ^{13}C NMR spectral parameters chosen for the analysis permitted quantitative assessment of the epoxide content of the oxidized CB to be made, using the area of the 56 ppm and 58 ppm resonances relative to the area of the saturated carbon resonances (25–33 ppm region). For this determination the ^{13}C NMR spectra of the soluble extracts were used instead of those of the corresponding gel samples, since the former displayed better resolution and higher signal-to-noise ratio than the latter. However, the spectra of the sol samples offered no qualitative information beyond that exhibited by Fig. 1 and so were omitted here for convenience. Use of the soluble extracts also permitted some additional features of the microstructure of oxidized CB to be noted. Broad resonances in the 65–77 ppm (alcohols) and 78–88 ppm (peroxides) regions could be observed in the ^{13}C NMR spectra. Unfortunately, accurate measurements of the intensities of these resonances could not be made. The infrared spectra of the extracts were in accord with published spectra for oxidized CB [5,6], showing a loss of —CH=CH— unsaturation and the growth of hydroxyl (2.9 μm) and carbonyl ($\sim$5.8 μm) bands, as well as generally enhanced absorption between 7–12 μm. The alcohol content of the oxidized CB samples was estimated from the A'_{OH} values assuming that the band at 2.9 μm was due solely to alcohol groups with no contribution from hydroperoxide groups. This assumption was justified by the observation made by Beavan and Phillips [6], and recently corroborated in this laboratory, that the intensity of the 2.9-μm band in the infrared spectrum of thermally oxidized CB does not decrease on exposure to ultraviolet light in vacuum. For a typical sample of CB oxidized at 125°C for 75 min (as in Fig. 1), and having 0.56 oxygen atom per monomer unit, the oxygen content comprised mainly epoxides and alcohols with 0.19 and 0.12 oxygen atoms per monomer unit, respectively.

Although epoxide bands [11] at 7.2, 7.9, and 11.1 μm could be discerned in the infrared spectra of the oxidized CB, those bands could not be used for quantitative assessment of the epoxide content. Neither was it possible to use the observed 2.9-ppm epoxide resonance [11] in the corresponding ^{1}H NMR spectra for quantitative purposes.

IV. DISCUSSION

The formation of epoxides in the thermal oxidation of simple olefins is well established [12]. Recently, the occurrence of epoxides in the thermal oxidation of 1,4-polyisoprene has been discussed [8,13]. However, the presence or importance of epoxides in the autoxidation of 1,4-polybutadiene has apparently not been previously recognized. As this study shows, not only are epoxides present in thermally oxidized CB, they constitute the major product. Their formation results from the following reaction steps, representing a variation on the accepted mechanism for autoxidation of unsaturated polymers [1–3]:

$$R \cdot \text{ (from initiation)} + O_2 \longrightarrow RO_2 \cdot \tag{1}$$

$$RO_2 \cdot + RH \longrightarrow RO_2H + R \cdot \tag{2}$$

$$RO_2H \longrightarrow RO \cdot + \cdot OH \tag{3}$$

$$2RO_2H \longrightarrow RO \cdot + RO_2 \cdot + H_2O \tag{4}$$

$$RO_2 + \underset{}{\overset{}{C}} = \underset{}{\overset{}{C} } \longrightarrow RO_2 - \underset{}{\overset{|}{C}} - \underset{}{\overset{|}{C}} \cdot \text{ (I)} \tag{5}$$

$$RO_2 - \underset{|}{\overset{|}{C}} - \underset{|}{\overset{|}{C}} \cdot + O_2 \longrightarrow RO_2 - \underset{|}{\overset{|}{C}} - \underset{|}{\overset{|}{C}} - O_2 \cdot \tag{6}$$

$$RO_2 - \underset{|}{\overset{|}{C}} - \underset{|}{\overset{|}{C}} \cdot \longrightarrow RO \cdot + \underset{\overset{\backslash}{\underset{O}{C} - C}}{} \tag{7}$$

The fact that epoxides constitute some 30–50% of the total product from the thermal oxidation of CB indicates that addition of a peroxy radical to a double bond to give the intermediate (I) (Reaction 5) is clearly more important than abstraction of hydrogen by that radical (Reaction 2). The absence of hydroperoxides in the thermally oxidized CB noted earlier requires that any RO_2H formed by Reaction 2 readily disappears through Reactions 3 and 4. The temporary crosslink, (I), may either capture oxygen to consolidate the crosslink (Reaction 6) or else dissociate to yield an alkoxy radical and an epoxide (Reaction 7). The partitioning between Reactions 6 and 7 must favor the latter, as shown by the preponderance of epoxides. The fate of the alkoxy radicals formed in Reaction 7 cannot be stated definitely at this time, although an important route presumably involves hydrogen abstraction to give alcohols [14]. Other reactions expected for the RO· radicals are addition to a double bond and β scission to give aldehydes and alkyl radicals [6,14]. The involvement of Reactions 5 and 7 is in accord with the formation of mainly trans epoxides from an all cis-1,4-polybutadiene, with rotation in the intermediate radical, (I), competing with dissociation to epoxide.

The question remains as to why epoxide formation is relatively more important in the autoxidation of CB than it is for 1,4-polyisoprene [8]. A plausible explanation may involve the expected stabilities of the radical, (I). While CB necessarily gives a secondary radical, 1,4-polyisoprene may give a more stable tertiary radical. The CB intermediate would therefore require less energy for the rearrangement to epoxide and, moreover, is less likely to capture oxygen due to its shorter lifetime. This latter effect is compounded by the fact that CB crosslinks during oxidation, becoming in effect more viscous and thereby reducing the rate of diffusion of oxygen into the polymer film.

Although the dialkyl peroxides formed by Reaction 6 constitute a small portion of the oxidation product from CB, such species are presumably responsible for the major alterations in the macroscopic properties of CB following oxidation. However, the methods used in this study are not sufficiently sensitive to assess the extent of crosslinking and chain scission.

V. SUMMARY

A study of the microstructural changes occuring in CB during thermal, uncatalyzed oxidation was carried out. Although the oxidation of CB is accompanied by extensive crosslinking with attendant insolubilization, it was found possible to follow the oxidation of solid CB directly with ^{13}C NMR spectroscopy. The predominant products appearing in the ^{13}C NMR spectra of oxidized CB are epoxides. The presence of lesser amounts of alcohols, peroxides, and carbonyl structures was adduced from complementary infrared and NMR spectra of soluble extracts obtained from the oxidized, crosslinked CB. This distribution of functional groups contrasts with that previously reported for the autoxidation of 1,4-polyisoprene. The difference was rationalized in terms of the relative stabilities of intermediate radical species involved in the autoxidation of CB and 1,4-polyisoprene.

REFERENCES

1. E. M. Bevilacqua, "Thermal Stability of Polymers" (R. T. Conley, ed.), Vol. 1, p. 189. Dekker, New York, 1970.
2. J. R. Shelton, *Rubber Chem. Technol.* **45**, 359 (1972).
3. J. A. Howard, *Rubber Chem. Technol.*, **47**, 976 (1974).
4. E. M. Bevilacqua, *J. Polym. Sci., Part C* **24**, 285 (1968).
5. R. L. Pecsok, P. C. Painter, J. R. Shelton, and J. L. Koenig, *Rubber Chem. Technol.* **49**, 1010 (1976).
6. S. W. Beavan and D. Phillips, *Eur. Polym. J.* **10**, 593 (1974); *Rubber Chem. Technol.* **18**, 692 (1975).
7. M. L. Kaplan and P. G. Kelleher, *Science* **169**, 1206 (1970); *J. Polym. Sci., Part A-1* **8**, 3163 (1970); *Rubber Chem. Technol.* **44**, 642 (1971); *ibid.* **45**, 423 (1972).
8. M. A. Golub, M. S. Hsu, and L. A. Wilson, *Rubber Chem. Technol.* **48**, 953 (1975).
9. M. W. Duch and D. M. Grant, *Macromolecules* **3**, 165 (1970).
10. M. A. Golub, R. V. Gemmer, and M. L. Rosenberg, *Amer. Chem. Soc., Div. Polym. Chem., Prepr.* **18**, 357 (1977); *Advan. Chem. Ser.*, in press.
11. R. V. Gemmer and M. A. Golub, *J. Polym. Sci., Polym. Chem. Ed.*, in press.
12. W. F. Brill and B. J. Barone, *J. Org. Chem.* **29**, 140 (1964).
13. D. Barnard, M. E. Cain, J. I. Cunneen, and T. H. Houseman, *Rubber Chem. Technol.* **45**, 381 (1972).
14. J. Kochi, "Free Radicals" (J. Kochi, ed.), Vol. 2, p. 677. Wiley, New York, 1973.

7

NMR and Infrared Study of
Photosensitized Oxidation of Polyisoprene

MORTON A. GOLUB, MARK L. ROSENBERG, and
*ROBERT V. GEMMER**

AMES RESEARCH CENTER
NATIONAL AERONAUTICS AND SPACE ADMINISTRATION
MOFFETT FIELD, CALIFORNIA

I. INTRODUCTION

The thermal oxidation of diene polymers has been studied extensively [1,2], and the corresponding photooxidation has also received considerable attention [3,4]. Recently, a strong interest has developed in the photosensitized oxidation of 1,4-polybutadiene and 1,4-polyisoprene [5–15]. This interest arose from the realization that molecular oxygen in its lowest excited state ($^1\Delta_g$ or singlet oxygen, $^1O_2$, with energy of 22.5 kcal/mol), which is extremely reactive towards olefinic compounds [16,17], is implicated in

* National Research Council–National Aeronautics and Space Administration Research Associate 1975–1977. Present address: American Cyanamid Company, Stamford, Connecticut.

the surface aging or oxidation of unsaturated polymeric materials exposed to air and sunlight [14]. While the photooxidation of the diene polymers is assumed to follow the same free radical chain mechanism as the thermal oxidation [3,4], the photosensitized oxidation, to the extent that it involves $^1O_2$, is considered to yield allylic hydroperoxides with shifted double bonds, according to an "ene"-type process [16,17]

$$-\overset{|}{\underset{H}{C_1}}=\overset{|}{C_2}-\overset{|}{C_3}- + {}^1O_2 \longrightarrow -\overset{|}{\underset{O_2H}{C_1}}-\overset{|}{C_2}=\overset{|}{C_3}-$$

Such a process was noted by Kaplan and Kelleher [9] for the $^1O_2$ reaction with squalene, a model compound for 1,4-polyisoprene. Although it has been suggested [6,9] that various double-bond shifts probably occur in the $^1O_2$-polyisoprene reaction, only one has been exhibited to date, namely, the shift to exomethylene groups in squalene [9]. The occurrence of other double-bond shifts accompanying hydroperoxidation of squalene (SQ), and indeed the occurrence of various shifts in the analogous reactions involving cis-1,4-polyisoprene (CI) and trans-1,4-polyisoprene (TI), needed to be demonstrated. Moreover, since cis and trans forms of 1,4-polybutadiene were reported to react with $^1O_2$ by different, though unspecified, mechanisms [8,9], the possibility that the isomeric forms of 1,4-polyisoprene likewise undergo different reactions with $^1O_2$ warranted examination.

Prior work on the structural analysis of diene polymers and SQ subjected to $^1O_2$ attack has focused on infrared (ir) spectroscopy, principally the 2.9-μm band as a qualitative indicator for hydroperoxides. As an extension of that work as well as a follow-up to our previous nuclear magnetic resonance (NMR) study of the thermal oxidation of polyisoprene [18], this paper presents the first investigation of the microstructural changes attending the photosensitized oxidation of CI, TI, and SQ, using ^1H and ^{13}C NMR as well as ir spectroscopy. In this paper we show that TI and SQ are alike in undergoing the three possible double bond shifts while CI undergoes readily only two of these shifts (to di- and trisubstituted double bonds) and relatively little of the third (to exomethylene groups). We also show that the ir absorbance ratio, $A_{2.9}/A_{6.9}$, provides a satisfactory quantitative measure of the extent of hydroperoxidation in the singlet oxygenation of CI, TI, or SQ.

II. EXPERIMENTAL TECHNIQUE

Purified samples of CI, TI, and SQ were similar to previous samples [18]. Aliquots (100–150 ml) of C_6H_6 solutions of CI or TI ($\sim$5 gm/l) or SQ ($\sim$10 gm/l), containing $\sim$10 mg of photosensitizer/gm substrate, were irradiated at $\sim 15°C$ under one atmosphere of O_2, in a magnetically stirred Pyrex cell placed on a cooling plate. The light source was a 35-mm slide projector with a Corning CS3-71 filter. The O_2 uptake was followed manometrically. The visible light sensitizers were commercial samples of oil-soluble chlorophyll, methylene blue, Rose Bengal, and acridine orange. Before addition to the substrate solution, the dyes were first dissolved in 5 ml CH_3OH, while chlorophyll was added directly. In addition, several photosensitized oxidations were also carried out with polymer-bound Rose Bengal [19] (PHOTOX^{TM}, Hydron Laboratories, Inc., New Brunswick, New Jersey) in suspension (0.3 gm insoluble beads per gm substrate).

After various O_2 uptakes, the products were analyzed by ir spectra run on thin polymer films (or viscous layers of $^1O_2$-reacted SQ) cast onto NaCl plates from the reaction solutions. The solutions were then rotary evaporated at room temperature, and the residues dissolved in $CDCl_3$ or C_6D_6 for ^1H or ^{13}C NMR analysis. Since the $^1O_2$-reacted CI tended to be unstable, more so than the corresponding products from TI or SQ, it was necessary to carry out the ir and NMR analyses promptly to avoid developing extraneous absorptions or resonances associated with subsequent autoxidation. In several experiments, portions of the hydroperoxidized solutions were also reduced to the corresponding alcohol forms with excess $NaBH_4$, followed by aqueous acidic workup and drying with $MgSO_4$ (in the case of SQ) or by neutralization with glacial acetic acid (in the case of CI or TI) and subsequent rotary evaporation and spectroscopic analysis.

Infrared spectra were obtained with a Perkin-Elmer Model 180 spectrometer. ^1H and ^{13}C NMR spectra were obtained with a Varian HA-100D and a Varian CFT-20 spectrometer, respectively, as described previously [18]. Chemical shifts are reported in ppm relative to tetramethylsilane.

III. RESULTS AND DISCUSSION

A. Infrared Spectroscopy

Figure 1 shows the ir changes associated with the photosensitized oxidation of CI, using chlorophyll as sensitizer. The major changes are the development of strong bands at 2.9 μm (OH, as hydroperoxides) and 10.3 μm (*trans* —CH=CH—), and the diminution of the 12.0-μm band (*cis* —C(CH$_3$)=CH—). The broad absorption at ~ 11.3 μm (C=CH$_2$) suggests the formation of some exomethylenes, but this cannot be important since the 6.0-μm peak (internal C=C) lacks a shoulder at 6.1 μm (external C=C). The ir data are indicative of a $^1O_2$–CI reaction leading to allylic hydroperoxides accompanied by double bond shifts to (I) and a minor amount of (III);

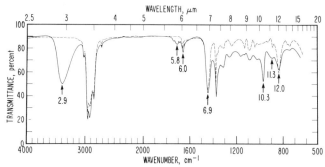

FIG. 1   Infrared spectra of CI before (–––) and after (——) singlet oxygenation with chlorophyll as sensitizer.

although not detectable by ir, a shift to (II) is revealed by ^{1}H NMR, as discussed below. The weak absorption observed at 5.8 μm is due to residual

$$\text{(I)} \qquad\qquad \text{(II)} \qquad\qquad \text{(III)}$$

chlorophyll in the cast film of ^{1}O$_2$-reacted CI, and not to carbonyl groups. In the case of photosensitization with methylene blue, no 5.8-μm peak is observed but instead an extraneous peak appears at 6.25 μm due to that dye. The latter peak can, however, be removed by washing the cast film with methanol. Otherwise, the ir spectra obtained using methylene blue or acridine orange as sensitizers were closely similar to that shown in Fig. 1 for chlorophyll sensitization. Rose Bengal, on the other hand, gave rise to additional absorptions at 9.1 and 9.9 μm (C—O—C), indicating that some sensitizer-induced autoxidation occurred alongside the singlet oxygenation.

   It should be stressed that the ir spectrum of ^{1}O$_2$-reacted CI typified by Fig. 1 is fundamentally different from the spectra of thermally or photochemically oxidized *cis*-1,4-polyisoprene [20,21]. The latter spectra show broad intense absorption throughout the 8- to 12-μm region (due to C—O groups) which mask the unsaturation bands of interest.

   A suitable measure of the degree of hydroperoxidation is afforded by the absorbance ratio, $A_{2.9}/A_{6.9} \equiv A'$, where the 6.9-$\mu$m band (CH$_2$ bending vibration) is an internal standard. Figure 1 with $A' = 0.76$ was obtained for an oxygen uptake of 0.27 O$_2$/monomer unit. As shown in Fig. 2, a smooth plot of A' versus oxygen uptake was obtained for CI using methylene blue and chlorophyll as photosensitizers. Data for Rose Bengal were erratic and tended to fall to the right of the curve due to autoxidative side reactions,

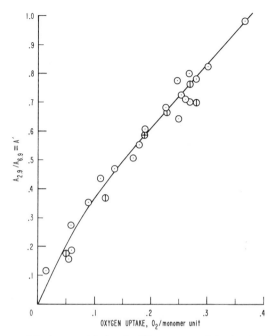

FIG. 2   Extent of hydroperoxidation in CI, represented by ir absorbance parameter, A', plotted versus oxygen uptake. $\bigcirc$, methylene blue; $\oplus$, chlorophyll; $\oplus$, tetraphenylporphine.

and were therefore omitted from Fig. 2. Likewise, data for acridine orange were omitted since they showed some deviation from the smooth plot in Fig. 2, for reasons that are not clear. Interestingly, the use of tetraphenylporphine as photosensitizer for the $^1O_2$–CI reaction (by Dr. G. D. Mendenhall, Battelle Memorial Institute, Columbus, Ohio) gave a data point that fell nicely on the A' plot shown. Similar plots were obtained for singlet oxygenation of TI and SQ, using methylene blue or chlorophyll.

That the A' plot represents quantitative incorporation of O_2 as hydroperoxide may be seen from the following argument. The A' plot for SQ was found to pass through a point with $A' = 0.55$ corresponding to SQ monohydroperoxide ($= 0.167$ O_2/monomer unit). Dividing this A' value by the factor 1.6 obtained in this work for $A'_{SQ–O_2H}/A'_{SQ–OH}$ yields an $A' \sim 0.34$ for SQ monoalcohol that is virtually the same as the value (~ 0.33) we estimate from published spectra [9]. The corresponding factor in the case of CI, i.e., $A'_{CI–O_2H}/A'_{CI–OH}$, was found to be ~ 1.4.

As shown in Fig. 3, TI and SQ (a hexaisoprene with trans internal double bonds) display virtually identical ir changes as a result of photosensitized oxidation with methylene blue. Similar spectra were obtained with chlorophyll and acridine orange, while Rose Bengal again gave indications of

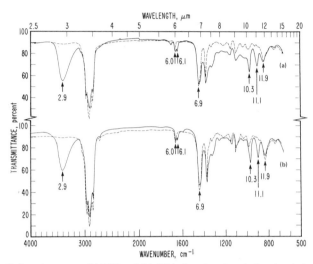

FIG. 3   Infrared spectra of (a) TI and (b) SQ before (– – –) and after (——) singlet oxygenation with methylene blue. The TI spectrum before reaction is that of amorphous balata.

autoxidative side reactions with TI but not SQ. The two trans isoprenic compounds (Fig. 3) not only show the development of the 2.9- and 10.3-μm bands and the diminution of the 11.9-μm band (*trans* —C(CH$_3$)=CH—), by analogy to CI, they also show significant absorption at 11.1 μm and 6.1 μm, indicative of exomethylene groups. The latter two bands were previously noted by Kaplan and Kelleher [9], and indeed spectrum B ($A' \sim 0.56$) may be regarded as the hydroperoxide counterpart of their SQ monoalcohol spectra ($A' \sim 0.33$). The ir changes observed here reveal that TI and SQ, in common with CI, yield structure (I), but, unlike CI, yield also a relatively large amount of (III), while ^{1}H NMR data below indicate that all three compounds yield (II). It should be mentioned that the main difference between the SQ monoalcohol spectra [9] and spectrum B, apart from the obvious difference in A' values, is that the latter has a much stronger 10.3-μm band, and hence contains a relatively larger amount of (I), than the former spectra. This may be due to the fact that spectrum B was obtained from the total, unchromatographed, hydroperoxidized SQ, whereas the cited spectra [9] represented a chromatographically separated product.

Before considering the other spectroscopic results, it should be noted that polymer-bound Rose Bengal, in common with the free Rose Bengal, yielded concurrent autoxidation and singlet oxygenation in the case of CI and TI. Moreover, the rates of O$_2$ uptake by these polyisoprenes using the polymer-bound dye were found to be much lower than the expected rates, assuming that the latter dye has an efficiency [19] of about 65% of that of the free dye. On the other hand, the free and polymer-bound dyes gave normal singlet oxygenation of SQ, uncomplicated by autoxidation and at comparable rates. Evidently, Rose Bengal, whether free or polymer bound, must react with the polyisoprenes in some manner to produce polymeric radicals which in turn give rise to peroxy radicals and ensuant autoxidation. The abnormally low efficiency of the polymer-bound dye in the case of the polyisoprenes, in contrast to that for SQ and low molecular weight olefins, presumably is due to the need for the substrate to diffuse into the insoluble sensitizer bead, which would be much more difficult for a macromolecule than for a small molecule.

B. Proton NMR Spectroscopy

Figure 4 shows the ^{1}H NMR spectrum of $^1O_2$-reacted CI corresponding to the ir spectrum in Fig. 1. New NMR resonances appear at 1.30

$(CH_3\overset{|}{C}\!\!-\!\!O\!\!-\!\!)$, 2.80 $(=\!\!\overset{|}{C}\!\!-\!\!CH_2\!\!-\!\!\overset{|}{C}\!\!=)$ [22], 4.20 $(-\!\!\overset{|}{C}H\!\!-\!\!O\!\!-\!\!)$, and

5.54 ppm $(-CH\!\!=\!\!CH\!\!-)$, while the original resonances at 2.05 ppm

$(-CH_2\overset{|}{C}\!\!=)$ and 5.12 ppm $(-\overset{|}{C}\!\!=\!\!CH\!\!-)$ are correspondingly reduced in intensity. Analogous NMR changes were observed for TI and SQ except that the latter compounds also showed new resonances at 4.8–4.9 ppm

$(\overset{\backslash}{\underset{/}{C}}\!\!=\!\!CH_2)$. Although not as obvious, a decrease is also observed in the

relative intensity of the 1.69-ppm resonance (cis $CH_3\overset{|}{C}\!\!=$) in CI, which has

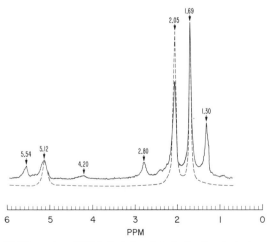

FIG. 4 ^{1}H NMR spectra of CI before (– – –) and after (——) photosensitized oxidation, corresponding to ir spectrum of Fig. 1.

counterparts at 1.62 ppm in TI, and at 1.61 ppm and 1.70 ppm (*trans* and

cis CH$_3$C$=$, respectively) in SQ [18]. The —O$_2$H protons could not be detected in the ^{1}H NMR spectra of ^{1}O$_2$-reacted CI but were barely detected as a broad resonance centered at $\sim$8 ppm in the spectra of ^{1}O$_2$-reacted TI and SQ. These ^{1}H NMR data not only corroborate the ir indications for double bond shifts (to (I) in CI, and to (I) and (III) in TI and SQ), they also establish the formation of (II) in all three compounds. Thus, the 5.54-ppm resonance is in accord with the disubstituted double bond in (I), the 4.20-ppm resonance represents the methine proton in (II) (or (III)), the 2.80-ppm resonance is ascribed to the "skipped" methylenes [22] resulting from double bond shift to (I) or (II) (as in (I′) or (II′)), while the 4.8–4.9 ppm resonance in TI or SQ signifies exomethylene groups in (III).

(I′)   (II′)

Using the relative areas of the 5.54- and 4.20-ppm resonances, and assuming a negligible amount of (III) in the ^{1}O$_2$-reacted CI depicted in Figs. 1 and 4, we estimated the 27% hydroperoxide content in this material to be divided almost equally between (I) and (II). For the TI sample shown in Fig. 3, which also had a 27% hydroperoxide content, we estimated using the corresponding ^{1}H resonances as well as the 10.3- and 11.1-μm ir bands, that (I), (II), and (III) were present in the relative amounts of 1.8:1.0:1.0. These results suggest that the cis and trans polymers have the same tendency to form (I), while the tendency to form (II) in the cis polymer is divided nearly equally between tendencies to form (II) and (III) in the trans polymer. Although there is considerable literature on the stereoselectivity of the ^{1}O$_2$– olefin reaction [16,17], apparently there are no comparable data for the exact mono-olefin analogs of CI and TI. Further work on such models would be desirable.

C. Carbon-13 NMR Spectroscopy

Typical ^{13}C spectra of ^{1}O$_2$-reacted CI, before and after reduction, are shown in Fig. 5. These spectra, which may be obtained with chlorophyll, methylene blue, or acridine orange as photosensitizers, but not Rose Bengal, are of interest on several counts. Firstly, spectrum A shows that the band of resonances at $\sim$80–93 ppm is quite different from the peroxide–hydro-peroxide–alcohol resonances at $\sim$70–90 ppm in the ^{13}C spectrum of thermally oxidized CI [18]. Rose Bengal, on the other hand, in line with

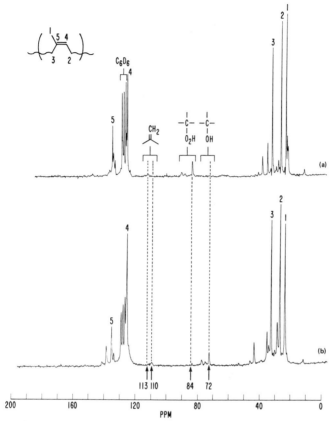

FIG. 5 ^{13}C NMR spectra of $^{1}O_{2}$-reacted CI (a) before and (b) after reduction with NaBH$_4$. Spectrum A corresponds to ir and ^{1}H NMR spectra of Figs. 1 and 4.

its propensity to induce autoxidation in CI, often produces a band of resonances at ~60–85 ppm that resembles the corresponding region in the ^{13}C spectrum of thermally oxidized CI. Secondly, Fig. 5 supports the arguments presented above that CI undergoes very little double bond shift to (III) compared to TI and SQ. This is evidenced by the fact that the 113–110 ppm resonances in the case of CI are markedly weaker than those in TI or SQ (Figs. 6 and 7). Thirdly, Fig. 5, along with Figs. 6 and 7, provides ^{13}C spectroscopic data on the effect of converting a hydroperoxide to an alcohol group in isoprenic compounds. Thus, there is an upfield shift of 3 ppm for the exomethylene carbon resonances in CI, TI, and SQ, and an upfield shift of 12 ppm in CI and 14 ppm in both TI and SQ for the oxygen-bearing carbon resonances, as a result of the reduction. Interestingly, similar upfield shifts

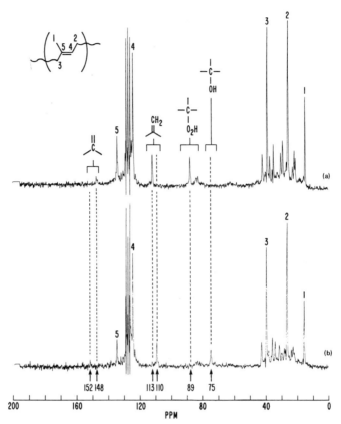

FIG. 6 ^{13}C NMR spectra of $^1O_2$-reacted TI (a) before and (b) after reduction with NaBH$_4$. Spectrum A corresponds to upper solid spectrum in Fig. 3.

of ∼ 12–13 ppm were observed by Olah and coworkers [23] for the —C—O—

resonances in various *tert*-alkyl hydroperoxides on reduction to the corresponding alcohols.

The ^{13}C spectra in Figs. 6 and 7, besides reinforcing the close similarity noted between the TI and SQ reactions with $^1O_2$, offer additional spectro-scopic data. The resonances at 113 ppm may be assigned to the exomethylene carbon in an internal isoprene unit, while the resonance at 114 ppm is assignable to the corresponding carbon in an external isoprene unit (as in SQ). Also, the resonance at 148 ppm, associated with the carbon attached to the exomethylene group, is shifted downfield by 4 ppm in both TI and SQ upon reducing the hydroperoxide to the alcohol.

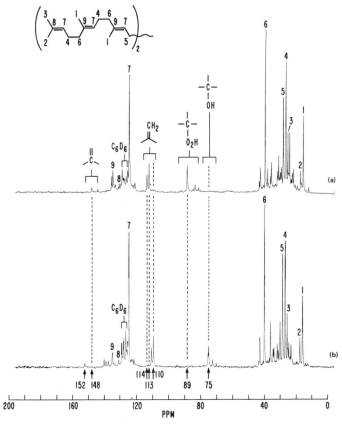

FIG. 7 ^{13}C NMR spectra of ^{1}O$_{2}$-reacted SQ (a) before and (b) after reduction with NaBH$_{4}$. Spectrum A corresponds to lower solid spectrum in Fig. 3.

IV. KINETICS OF SINGLET OXYGENATION

Finally, the photosensitized oxidation of CI, TI, and SQ using various visible light sensitizers gave the expected zero-order kinetic plots for oxygen uptake characteristic of singlet oxygenation. From the slopes of the straight line kinetic plots obtained for a number of runs with CI and TI, carried out under identical conditions and using methylene blue as sensitizer, the relative rates for CI and TI were found to be approximately 1.0:1.5. This ratio is remarkably close to those (1.0:1.3–1.5) indicated for the ^{1}O$_{2}$ reactions with *cis*- and *trans*-3-methyl-2-pentene, respectively [24], compounds which may be considered here as models for the polyisoprenes. It is also noteworthy that the photosensitized oxidation of the isoprenic compounds is quenchable with 1,4-diazabicyclo [2,2,2]octane (DABCO), a well-known quencher [25] of $^1O_2$.

V. SUMMARY

The microstructural changes which occur in *cis*- and *trans*-1,4-poly-isoprenes and in squalene during photosensitized oxidation were investigated with the aid of infrared and proton and carbon-13 NMR spectroscopy. The singlet oxygenation of these isoprenic compounds resulted in allylic hydro-peroxides with shifted double bonds, according to the expected "ene"-type process. In contrast to *trans*-1,4-polyisoprene and squalene, which displayed the three possible double bond shifts, *cis*-1,4-polyisoprene showed essentially two of the shifts (to di- and trisubstituted double bonds) and very little of the third (to exomethylene groups). A suitable measure of the extent of hydro-peroxidation was afforded by the absorbance ratio, $A_{2.9}/A_{6.9} \equiv A'$. Similar correlations of A' with oxygen uptake were obtained for the three isoprenic compounds, using chlorophyll or methylene blue as sensitizer. The use of Rose Bengal gave erratic results indicative of some autoxidation accom-panying the hydroperoxide formation. The singlet oxygenation followed zero-order kinetics, the relative rates for *cis*- and *trans*-1,4-polyisoprenes being approximately 1.0:1.5.

ACKNOWLEDGMENTS

The authors are grateful to Dr. Ming–ta S. Hsu for providing the ^{1}H NMR spectra, and to Dr. G. D. Mendenhall, Battelle–Columbus, for communicating pertinent data on tetraphenyl-porphine-photosensitized oxidation of *cis*-1,4-polyisoprene.

REFERENCES

1. J. R. Shelton, *Rubber Chem. Technol.* **45**, 359 (1972).
2. J. A. Howard, *Rubber Chem. Technol.* **47**, 976 (1974).
3. J. L. Morand, *Rubber Chem. Technol.* **47**, 1094 (1974).
4. B. Rånby and J. F. Rabek, "Photodegradation, Photooxidation and Photostabilization of Polymers," pp. 206–210. Wiley, New York, 1975.
5. T. Mill, K C. Irwin, and F. R. Mayo, *Rubber Chem. Technol.* **41**, 296 (1968).
6. J. F. Rabek, *XXIII IUPAC Congress, Boston, USA, 1971*, Vol. 8, p. 29. Butterworth, London, 1971.
7. M. L. Kaplan and P. G. Kelleher, *Science* **169**, 1206 (1970).
8. M. L. Kaplan and P. G. Kelleher, *J. Polym. Sci., Part A-1* **8**, 3163 (1970); *Rubber Chem. Technol.* **44**, 642 (1971).
9. M. L. Kaplan and P. G. Kelleher, *Rubber Chem. Technol.* **45**, 423 (1972).
10. G. P. Canva and J. J. Canva, *Rubber J.* **153**, 36 (1971).
11. A. K. Breck, C. L. Taylor, K. E. Russell, and J. K. S. Wan, *J. Polym. Sci., Polym. Chem. Ed.* **12**, 1505 (1974).

12. A. Zweig and W. A. Henderson, Jr., *J. Polym. Sci., Polym. Chem. Ed.* **13**, 717, 993 (1975).
13. B. Rånby and J. F. Rabek, "Photodegradation, Photooxidation and Photostabilization of Polymers," pp. 274–5, 315–326. Wiley, New York, 1975.
14. D. J. Carlsson and D. M. Wiles, *Rubber Chem. Technol.* **47**, 991 (1974).
15. J. F. Rabek and B. Rånby, *J. Polym. Sci., Polym. Chem. Ed.* **14**, 1463 (1976).
16. D. R. Kearns, *Chem. Rev.* **71**, 395 (1971).
17. C. S. Foote, *Pure Appl. Chem.* **27**, 635 (1971).
18. M. A. Golub, M. S. Hsu, and L. A. Wilson, *Rubber Chem Technol.* **48**, 953 (1975).
19. A. P. Schaap, A. L. Thayer, E. C. Blossey, and D. C. Neckers, *J. Amer. Chem. Soc.* **97**, 3741 (1975).
20. G. Salomon and A. C. Van der Schee, *J. Polym. Sci.* **14**, 181 (1954).
21. A Tkáč and V. Kellö, *Rubber Chem. Technol.* **28**, 383 (1955).
22. M. van Gorkom and G. E. Hall, *Spectrochim. Acta* **22**, 990 (1966).
23. G. A. Olah, D. G. Parker, N. Yoneda and F. Pelizza, *J. Amer. Chem. Soc.* **98**, 2245 (1976).
24. R. Higgins, C. S. Foote, and H. Cheng, *Advan. Chem. Ser.* **77**, 102 (1968).
25. C. Quannès and T. Wilson, *J. Amer. Chem. Soc.* **90**, 6527 (1968).

8

Chemiluminescence Study of the Autoxidation of *cis*-1,4-Polyisoprene

G. DAVID MENDENHALL and RICHARD A. NATHAN

BATTELLE LABORATORIES
COLUMBUS, OHIO

MORTON A. GOLUB

AMES RESEARCH CENTER
NATIONAL AERONAUTICS AND SPACE ADMINISTRATION
MOFFETT FIELD, CALIFORNIA

I. INTRODUCTION

The free-radical mechanism for the autoxidation of *cis*-1,4-polyisoprene (natural rubber or its synthetic counterpart) has been investigated extensively [1]. An important feature of this mechanism, and indeed also of the autoxidation of hydrocarbons generally [2], is that it is a chain process propagated by alkyl and peroxy radicals and terminated through bimolecular reactions involving these same radicals. In the usual oxidation situation, that is, at all oxygen pressures greater than a few torr, the alkyl radicals are rapidly converted to peroxy radicals, and the termination step proceeds almost exclusively through the latter radicals. The bimolecular decay of the peroxy radicals is accompanied by a weak emission of light or chemiluminescence [3]. Kinetic evidence is consistent with an electronically excited ketone produced in the termination reaction as the source of the emission [3].

The first observation of chemiluminescence from the oxidative degradation of polymers was reported by Ashby [4], who dealt mainly with polypropylene but made passing mention of several other polymers. Subsequently, a number of papers have appeared dealing with oxidative chemiluminescence from a variety of polymers [5]. In this paper we report the first detailed study of the chemiluminescence emitted in the autoxidation of *cis*-1,4-polyisoprene.

The chemiluminescence technique is extremely sensitive and can follow rates of oxidation that are too slow to be measured conveniently by other means [6]. This work thus offered the potential of throwing new light on the autoxidation of *cis*-1,4-polyisoprene, especially in the very early stages or under ambient conditions where conventional spectroscopic procedures are rather insensitive.

II. EXPERIMENTAL TECHNIQUE

A. Materials

Commercial-grade *cis*-1,4-polyisoprene (Ameripol SN), obtained from the B. F. Goodrich Research and Development Center, Brecksville, Ohio, was purified by three reprecipitations from benzene solution with methanol as precipitant. The solvents were reagent grade and were degassed by flushing with argon before use. Operations with the solutions were carried out in an argon or nitrogen atmosphere, and the final benzene stock solution (~15 gm/1) was stored in the dark under positive argon pressure. Polyisoprene films were prepared by spreading layers of stock solution over plate glass sheets and allowing the solvent to evaporate. The glass (2-mm thick, single-strength sheet) was scratched on the reverse side to facilitate breaking into rectangular samples (3.7 cm × 3.5 cm cross section). The glass-supported films were stored in argon and mounted on 35-mm aluminum slide holders prior to chemiluminescence experiments. The exposed area of the mounted slides was 8.4 cm^2 and the average exposed weight was 0.0042 gm. The films showed no noticeable differences in properties up to several months after preparation.

B. Apparatus and Procedure

A schematic view of the chemiluminescence apparatus is presented in Fig. 1. The heart of the instrument was a 600-W oven (with accessories for cooling) surrounded by a water-cooled shield to prevent the entire light-tight box from reaching the oven temperature. The oven was equipped for accurate variable temperature control (Foxboro Company heater-proportional control unit) and with gas inlets for introducing any desired atmosphere. The mounted polymer films were placed on the stainless steel

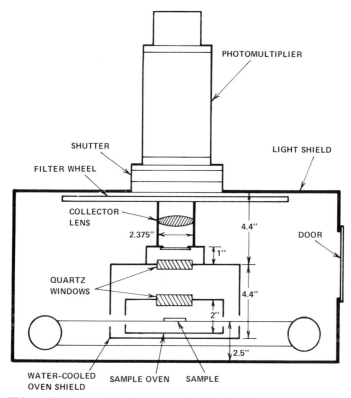

FIG. 1 Photon counting instrumentation for chemiluminescence studies.

conveyor belt, then passed through a double set of pressure-activated doors into the sample chamber. The correct position of the film samples was determined by appropriate marks on the conveyor belt.

Chemiluminescence emission from the samples was focused with a 2-inch quartz lens onto the detector of a 12-stage photomultiplier (RCA Model 4501/V4) operated at 1500 V. The photomultiplier output was connected to a counter (General Radio Model 1191) whose output, expressed in analog or digital form, was usually set to average the photon counts over a 10-sec interval. In this way, spurious signals due to high-frequency noise in the line voltage (appearing when room lights were on or from unidentified sources) were eliminated. A strip-chart recorder connected to the counter permitted automatic data collection. A wheel containing up to 20 filters was positioned in the optical path to permit spectral analysis of the chemiluminescence by measuring the relative emission intensity after the light passed through the various filters. A manual and an automatic shutter were also positioned in the optical path to permit background counting.

For chemiluminescence experiments below 117°C, the polyisoprene film, initially at 25°C, was placed in the oven preheated to the desired temperature. For experiments above 117°C, the time to warm the film to the oven temperature introduced a significant distortion in the shape of the chemiluminescence intensity–time plots. To circumvent this difficulty, the film was placed in the oven at 25°C and flushed with argon for > 20 min, and the oven was then heated to the desired temperature. At the start of the chemiluminescence measurement, the argon stream was changed to oxygen.

Data for the Arrhenius plots for low-temperature chemiluminescence from polymer films containing 9,10-diphenylanthracene or tetraphenylporphine were obtained from a simpler apparatus consisting of an aluminum box 10″ on a side, with provisions for attachment of the same photomultiplier tube. The temperature of the film was monitored with a thermocouple directly in contact with it. The low-intensity chemiluminescence from partially oxidized polyisoprene films (as well as from hydroperoxidized samples obtained through prior singlet oxygenation) was corrected for background emission. The latter signal, amounting to 30–50 counts/sec under the experimental conditions employed, remained stable during 6 months' operation and was believed to be due to stray sources of light originating in the oven and photomultiplier housing, inherent noise from the photomultiplier tube, and cosmic radiation.

C. Reaction with Singlet Molecular Oxygen ($^1O_2$)

A solution of 0.43 gm of purified cis-1,4-polyisoprene in 40 ml of benzene, 5.8×10^{-6} M in meso-tetraphenylporphine (Strem Chemicals), was placed in a Pyrex flask with provision for magnetic stirring and flushed with oxygen for 2 min. The flask was connected to a gas burette filled with oxygen, and the stirred mixture was irradiated at 6°C for 75 min with the output from a 500-W projector lamp filtered through a Corning CS3-71 cutoff filter. During this time the solution absorbed 29.2 ± 0.5 ml of oxygen, corresponding to about one hydroperoxide group per five monomer units. Absorption of oxygen by the solution was negligible before and after irradiation. The solution was used immediately for casting films for chemiluminescence measurements and infrared analysis. The infrared spectrum of the $^1O_2$-treated polyisoprene showed an O–H band at 3400 cm^{-1} and a weak C–O absorption at 1000 cm^{-1}, and was comparable to that of singlet oxygenated *cis*-1,4-polyisoprene shown elsewhere [7].

D. Oxygen Uptake

For the oxygen uptake measurements, a benzene solution containing 0.0834 gm of *cis*-1,4-polyisoprene was placed in a 50-ml flask, with a flattened bottom 3 cm in diameter. The solvent was removed by evacuation with an oil pump through a trap at $-78°C$. The evacuated flask was placed in a conventional oil bath at $117.6 \pm 0.2°C$, allowed to equilibrate, and oxygen was admitted from a rubber gas reservoir. The uptake of oxygen by the polymer was then followed with a mercury-filled burette. A blank run without polymer was carried out to correct for volume expansion due to heating of the oxygen on admission to the flask. The amounts of oxygen absorbed by the polymer after 10, 30, 60, and 90 min were 3.6, 12.5, 18.1, and 20.8 cm^3, respectively, corresponding to 0.12, 0.41, 0.60, and 0.69 oxygen molecules per monomer.

III. RESULTS

A. Chemiluminescence–Autoxidation

Representative curves obtained for the oxidative chemiluminescence from *cis*-1,4-polyisoprene are shown in Fig. 2. The parameters used to characterize these curves are I_{max}, the maximum intensity of chemiluminescence,* and $t_{1/n}$, the time required to reach $(1/n)I_{max}$. These parameters, in the ideal case, are related to rate constants for the autoxidation and excited state reaction sequence.

The initiation steps in the autoxidation may be bi- or unimolecular, or pseudounimolecular. In the bimolecular case, Scheme (I) is assumed to hold [3,8].

---

*The intensity values in Fig. 2 were those indicated on the chemiluminescence recorder and were not corrected for a geometrical factor (~ 150) and for photomultiplier counting efficiency ($15 \pm 10\%$, between 300–500 nm), which is a function of the emission wavelength.

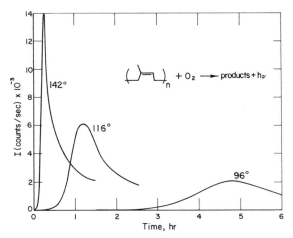

FIG. 2 Typical plots of chemiluminescence intensity as a function of time for *cis*-1,4-polyisoprene films in oxygen at several temperatures.

Scheme (I)

$$2RO_2H \rightleftharpoons (RO_2H)_2 \tag{1}$$

$$(RO_2H)_2 \longrightarrow RO\cdot + RO_2\cdot + H_2O \tag{2}$$

$$RO\cdot + RH \longrightarrow ROH + R\cdot \tag{3}$$

$$R\cdot + O_2 \longrightarrow RO_2\cdot \tag{4}$$

$$RO_2\cdot + RH \longrightarrow RO_2H + R\cdot \tag{5}$$

$$2RO_2\cdot \longrightarrow \text{nonradical products} + O_2 \tag{6}$$

$$2RO_2\cdot \longrightarrow P^* \tag{7}$$

$$P^* \longrightarrow P + h\nu \tag{8}$$

$$P^* \longrightarrow P \tag{9}$$

In this scheme, P^* is the electronically excited product responsible for chemiluminescence, RH is the polymer, and R denotes the different radical sites on the polyisoprene backbone. Reaction (9) designates the principal nonradiative decay pathway, which may be intersystems crossing, quenching by oxygen, or other reactions.

For unimolecular initiation the possible reactions include

$$O_2 + RH \longrightarrow R\cdot + HO_2\cdot \tag{10}$$

$$RO_2H \longrightarrow RO\cdot + \cdot OH \tag{11}$$

$$RO_2H + M \longrightarrow \longrightarrow 2R\cdot \tag{12}$$

where M is a metal ion or some functional group at constant concentration.

In an earlier version of this paper that we presented at the 172nd ACS meeting in San Francisco, we attempted to distinguish between unimolecular and bimolecular initiation based on equations derived from each case that were applied to the chemiluminescence data. This procedure involved severe approximations, since the polymer oxidation was extensive and the wavelengths of emission changed as the reaction proceeded (see below).

For a description of the time and temperature dependence of the chemiluminescence, the most convenient equations for the high-temperature data are (13) and (14). Equation (13) is derived by generalizing one developed by Lundeen and Livingston [8a] and based on earlier ones given by Tobolsky, Metz, and Mesrobian [8b] for tetralin autoxidation. Equation (13) is based on a unimolecular mode of initiation, and long free-radical chains, which may not obtain in practice. The data nevertheless can be fitted satisfactorily to it for $n < 20$ over the experimental temperature range. For this purpose we rearranged Eq. (13) and assumed that $[RO_2H]_0 \simeq 0$, giving Eq. (15).

$$k_{12} = A_{12}e^{-E_{12}/RT} = \frac{2}{t_{1/n}}\ln\left[\frac{1 - [(RO_2H)_0/(RO_2H)_\infty]^{1/2}}{1 - (1/n)^{1/2}}\right] \tag{13}$$

$$I_{max} = q_1 k_5^2[RH]^2/k_6 \qquad \text{where} \qquad q_1 = k_7 k_8/(k_8 + k_9) \tag{14}$$

$$\ln t_{1/n} = E_{12}/RT - \ln A_{12}' \qquad \text{where} \qquad A_{12}' = A_{12}/2\ln\left[\frac{1}{1 - (1/n)^{1/2}}\right] \tag{15}$$

A least-squares analysis of the data was conducted for plots of $\ln t_{1/n}$ vs. T^{-1} for different n. The resulting values of E_{12}, A_{12}', and A_{12} (except for $n = 1$) appear in Table I. Representative plots for two values of n appear in Fig. 3. It is noteworthy that the data when analyzed according to Eq. (15) give satisfactory correlation coefficients for $n = 1$, for which value the equation

TABLE I

Arrhenius Parameters for Chemiluminescence from cis-1,4-Polyisoprene according to Eq. (15)

n	E_{12}(kcal/mol)	$\ln A_{12}'$(min^{-1})	$\ln A_{12}$(min^{-1})	N	r
20	25.2 ± 4.7	30.0 ± 6.1	29.3 ± 6.1	14	0.976
10	26.4 ± 5.7	$30.9 + 7.4$	30.6 ± 7.4	20	0.980
4	24.4 ± 3.3	27.7 ± 4.2	28.0 ± 4.2	19	0.987
2	23.2 ± 2.2	25.4 ± 2.9	26.3 ± 2.9	21	0.991
$\frac{4}{3}$	22.4 ± 2.0	23.6 ± 2.6	25.0 ± 2.6	19	0.990
1	20.6 ± 1.5	—^{a}	22.5 ± 2.0	20	0.991
$\frac{4'}{3}$	18.8 ± 0.8	18.5 ± 1.0	19.9 ± 1.0	15	0.993
$2'$	17.1 ± 1.0	16.5 ± 1.4	17.4 ± 1.4	12	0.987

^{a} Not defined.

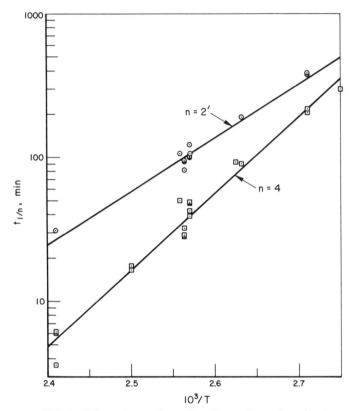

FIG. 3   Values of $t_{1/4}$ and $t_{1/2}$, plotted according to Reaction 5.

is discontinuous, and also for data obtained after the chemiluminescence maximum (indicated by prime signs). The scatter in the plots is believed to be due to errors in temperature measurement, since at a given oven setting the $t_{1/n}$ values were identical for duplicate runs or with twice the sample loading. From Eq. (14) we expect a linear relationship between ln I_{max} and T^{-1}. An Arrhenius plot for I_{max} is presented in Fig. 4, from which we obtain

$$E_{I_{max}} = 16.19 \pm 0.69 \text{ kcal/mol} \simeq 2E_5 - E_6$$

and

$$\log A_{I_{max}} = 18.2 \pm 0.40 \text{ photons/gm sec.}$$

This value of $E_{I_{max}}$ compares favorably with a value of $2E_5 - E_6 = 15$ kcal/mol estimated from oxygen-uptake measurements on natural rubber [8b].

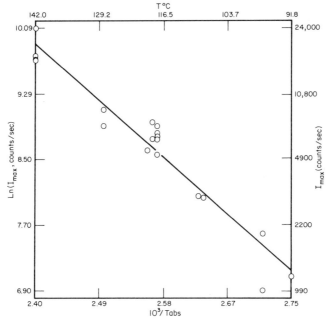

FIG. 4   Arrhenius plot of maximum chemiluminescence intensity from *cis*-1,4-polyisoprene.

For a bimolecular mode of initiation (Scheme (I)) an equation analogous to (14) also obtains with an additional factor of 4 in the numerator. The plot in Fig. 4 also shows scatter, probably from variations in film thickness and from irregularities in positioning of the sample in the oven of the apparatus.

B. Temperature-Jump Measurements

A film of *cis*-1,4-polyisoprene was allowed to oxidize at 118°C until the chemiluminescence reached a maximum intensity, at which point the film was cooled rapidly to 28°C. The steady-state emission was measured at this and several higher temperatures up to 66°C. The temperature was again lowered to 28°C and the procedure repeated (Fig. 5). Under these low-temperature conditions, the amount of initiator* and the chemiluminescence intensity do not change appreciably during a given measurement. In

---

* Formed in the autoxidation at 118°, along with other undetermined oxidation products.

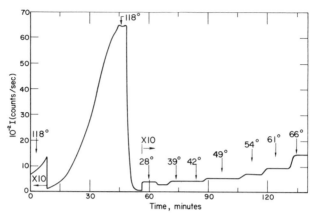

FIG. 5   Temperature-jump experiments with *cis*-1,4-polyisoprene in oxygen.

terms of the two modes of initiation, the following equations can then be obtained.

Unimolecular:

$$R_{init} = R_{term}$$
$$d[RO_2H]/dt = k_{12}[RO_2H] = k_6[RO_2\cdot]^2$$
$$I = d(h\nu)/dt = k_7 k_8 [RO_2\cdot]^2/(k_8 + k_9)$$
$$\cong k_7 k_8 k_{12}[RO_2H]/(k_6(k_8 + k_9)) \tag{16}$$
$$E_I \cong E_7 + E_8 + E_{12} - E_6 - E_9 \simeq E_{12} \qquad \text{for } k_9 \gg k_8 \tag{17}$$

Bimolecular:

$$k_2[(RO_2H)_2] = k_6[RO_2\cdot]^2$$
$$I \cong k_7 k_8 k_2 [(RO_2H)_2]/(k_6(k_8 + k_9))$$
$$\cong k_1 k_2 k_7 k_8 [(RO_2H)^2]/(k_{-1} k_6(k_8 + k_9)) \tag{18}$$
$$E_I \cong E_1 + E_2 + E_7 + E_8 - E_{-1} - E_6 - E_9 \cong E_1 + E_2 - E_{-1} \tag{19}$$

From the Arrhenius plot for the temperature-jump approach, E_I was found to be 19.8 kcal/mol. The low intensity of the emission prompted us to repeat the experiment, first with added 9,10-diphenylanthracene (0.5%) and then with a larger amount of oxidized material. These experiments gave similar and more precise values for E_a of 25.3 and 24.7 kcal/mol (Fig. 6).

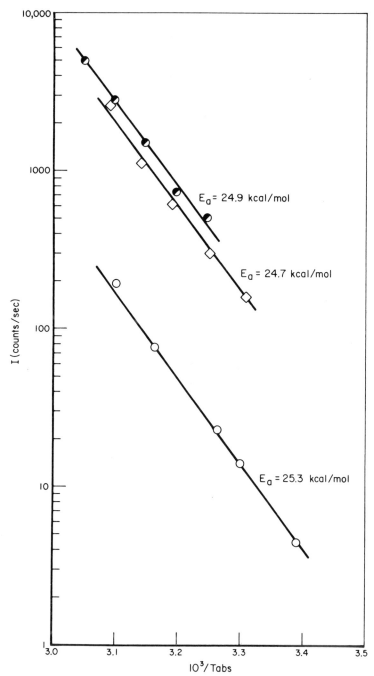

FIG. 6   Arrhenius plots for low-temperature chemiluminescence from *cis*-1,4-polyisoprene oxidized by different methods (○), autoxidized in oxygen at 118°C, dissolved in benzene and film recast with 0.5% 9,10-diphenylanthracene; (◇), 0.13 gm polyisoprene photooxidized with tetraphenylporphine sensitizer; (◕), 0.14 gm polyisoprene autoxidized 65 min at 98°C in air.

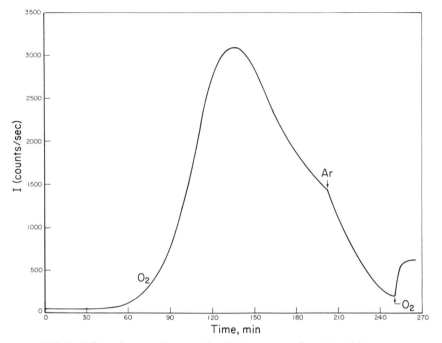

FIG. 7   Effect of atmosphere on chemiluminescence of *cis*-1,4-polyisoprene.

C. Chemiluminescence under Argon

As indicated in Fig. 7, when oxygen above an oxidizing sample of Ameripol SN was replaced by argon there was a sharp drop in the chemiluminescence intensity. However, even after about 1 hr in an inert atmosphere the intensity did not fall to the relatively low emission intensity (100 counts/sec) observed when the sample was placed initially under argon. Subsequent replacement of argon by oxygen caused the chemiluminescence to increase approximately to the level it would have had without the argon treatment.

D. Chemiluminescence from Singlet Oxygenated Polyisoprene

A sample of *cis*-1,4-polyisoprene containing about one hydroperoxide group per 5 monomer units was prepared via Eq. (20) (analog of the well-known "ene" reaction between singlet oxygen and olefins with allylic hydrogens [9,10]).

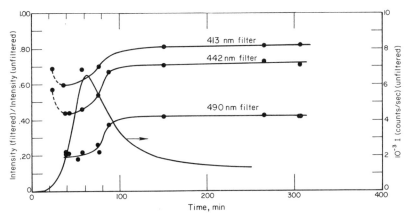

FIG. 8   Spectral distribution of chemiluminescence from cis-1,4-polyisoprene at 118°C in oxygen, determined with cutoff filters with O.D. = 0.50 at indicated wavelength.

The sample was heated from 25°C to 55°C during which time the chemiluminescence emission was monitored. The Arrhenius plot of chemiluminescence from the film (Fig. 8) yielded an activation energy of 24.9 kcal/mol, which may be taken as an alternative measure of E_I.*

E.   Special Distribution of Chemiluminescence

Although we were unable to obtain a spectrum of the weak chemiluminescence from cis-1,4-polyisoprene in a commercial spectrophotometer, some idea of its spectral distribution may be obtained with use of cutoff filters, as shown in Fig. 8. The emission at 118°C shown in the figure is seen to undergo a net shift first towards shorter wavelengths and then to longer wavelengths, and to display a nearly constant distribution of wavelengths after about 100 min. The shifts in spectral distribution to longer wavelengths with time of oxidation could not be due to self-absorption of the chemiluminescence at shorter wavelengths by colored materials formed during the autoxidation process, since the samples remained colorless.[†]

F.   Oxygen Uptake in Polymer

We measured the absorption of oxygen by a sample of cis-1,4-polyisoprene at 117°C, which was around the mean temperature of the chemiluminescence experiments. From the curve of oxygen absorbed versus time, we

---

* The reason is that the temperature dependence of chemiluminescence from the hydroperoxidized cis-1,4-polyisoprene has the same expression as that for the autoxidized material.

† Thus, a film with a thickness (0.0178 cm) twice that of most of the films studied here showed an increase in optical density of 0.02 between 400–550 nm after 75 min in oxygen at 117°C. For such a film it was calculated that the maximum loss in intensity due to an absorber was about 1% of the observed emission.

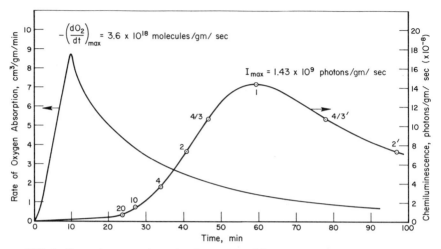

FIG. 9   Rate of oxygen absorption by *cis*-1,4-polyisoprene at 117°C and chemilumine-scence emission calculated from parameters from Eqs. (14) and (15). The values of *n* are indicated on the latter curve.

calculated the rate of absorption on a point-by-point basis (Fig. 9). The latter curve went through a maximum about 10 min after admission of oxygen and then declined in an exponential fashion. With a polymer sample twice as large the total amount of oxygen absorbed in 1.5 hr was only 1.4 times as much. The shape of the derivative curve, analogous to that in Fig. 9, was very similar and maximum rate of oxygen absorption appeared at the same time as in the first run. In both experiments small amounts of volatile products condensed on the cooler portions of the flask that contained the oxidizing polymer.

G. Chemiluminescence Efficiencies

Since there is no simple way to single out the particular reaction in the oxidizing polymer responsible for the chemiluminescence, an estimate of the latter's efficiency, a sort of reverse quantum yield, has to be based on some general feature of the overall process. One approach is based on the chemi-luminescence* from the singlet oxygenated polyisoprene. A 7.9-mg sample containing 1.6×10^{-5} mol RO_2H, displayed a constant emission of 1.1×10^6 photons/sec at 57°C. Steps (1) and (2) in the above reaction scheme may be considered as rate determining, while the concentration of hydroperoxide

---

* Estimated using the area under an idealized curve constructed from data in Table I and Fig. 5, and an assumed photomultiplier efficiency of 0.15 ± 0.10.

may be assumed constant during the time of observation. Walling and Heaton [11] studied the dimerization of *tert*-butyl hydroperoxide in CCl_4 and found $K_{eq} = \exp(-18.4/R + 5.95/RT)$. If this relation holds for the hydroperoxidized polymer, and one assumes the Arrhenius parameter $A_2 \simeq 10^{15 \pm 1}$ sec^{-1}, which corresponds to the value for simple alkyl peroxide decompositions [12], and that $E_1 - E_{-1} + E_2 = 22 \pm 2$ kcal/mol, then the rate of initiation is given by

$$R_{init} = e^{-18.4/R} \cdot 10^{15} e^{-(22 \pm 2/RT)} [RO_2H]^2$$

$$= 10^{-3.6 \pm 1.5} \quad M^{-1} \text{ sec}^{-1} [RO_2H]^2 \text{ at } 57°C$$

In our hydroperoxidized sample we estimate $[RO_2H] = 1.8$ M, and hence that $10^{15.6 \pm 1.5}$ hydroperoxide groups decompose each second. Dividing by the rate of photon emission, we obtain an efficiency of $10^{9.6 \pm 1.5}$ hydroperoxide groups decomposed per photon. This value is in satisfactory agreement with values of $10^{7.4}$ to $10^{9.8}$ initiating pairs per photon found by Kellogg [3d] for the chemiluminescence arising from radical-initiated oxidation of a series of liquid hydrocarbons.

We did not attempt correlations between chemiluminescence and oxygen uptake data at 117°C because the maximum chemiluminescence appeared long after the maximum rate of oxygen absorption had passed.

IV. DISCUSSION

A. Excitation Mechanism for Chemiluminescence

As disclosed in this work, chemiluminescence can arise from *cis*-1,4-polyisoprene by oxygen-dependent or oxygen-independent pathways. The intensity of chemiluminescence in an inert atmosphere was seen to vary from one polymer preparation to another, and to be greatly enhanced when the polymer was previously oxidized at elevated temperatures (Fig. 7). These observations suggest that peroxides and hydroperoxides, present originally in the films to varying extents, and produced in much larger amounts as a result of thermal autooxidation, can decompose in argon to yield the weak chemiluminescence noted. As for the more pronounced chemiluminescence in oxygen, it is generally assumed [3] that bimolecular termination involving peroxy radicals leads to the excited species responsible for the emission. Indeed, the disproportionation of alkylperoxy radicals has been suggested as the source of both excited triplet ketones and singlet molecular oxygen (3a,3d,13):

$$R_2CHO_2\cdot + R'O_2\cdot \rightleftarrows R_2CHO_2{-}O_2R' \longrightarrow R_2CO + O_2 + R'OH \qquad (21)$$

Another possible source of excited states, β scission of alkoxy radicals, may be dismissed as unlikely, since the process is nearly thermoneutral for those scissions expected in the polyisoprene system.

Since the chemiluminescence intensities reported in this study were very weak ($< 10^{10}$ photons/gm sec in the autoxidizing polymer and $\lesssim 10^8$ photons/gm sec in the hydroperoxidized polymer), it is possible that the light emission is due to an excited species occurring in a reaction which plays only a minor role in the oxidation mechanism. As an example, alkoxy radicals (which may be formed in a variety of ways in the oxidation of 1,4-polyisoprene [1,2,14]) undergo many metathesis reactions sufficiently exothermic to produce excited states:

$$(CH_3)_2CHO\cdot + R'\cdot \longrightarrow (CH_3)_2C{=}O + R'H \tag{22}$$

When $R'\cdot = RO\cdot$ or $RO_2\cdot$ in Reaction 5, $\Delta H_r = 89$ or 77 kcal/mol, respectively, which is large enough to permit excitation of acetone to its triplet state (78 kcal/mol) in the former case.

While the nature of the emitters in the autoxidation of cis-1,4-polyisoprene is not known, the high wavelengths in the spectral distribution do not correspond to simple ketones. Recently, Beaven and Phillips [15] recorded phosphorescence spectra from oxidized 1,4-polybutadiene films, at liquid nitrogen temperatures, and assigned the emission to α,β-unsaturated ketones. Interestingly, their phosphorescence wavelengths were close to those observed for chemiluminescence from cis-1,4-polyisoprene. Although infrared spectra of several polyisoprene films autoxidized on silver chloride sheets revealed no characteristic absorption at 1650 cm^{-1}, the presence of α,β-unsaturated ketones in these films could not be ruled out since the amount necessary to account for the observed chemiluminescence is well below the limit of infrared (ir) detection. We were unable also to record phosphorescence spectra from partially oxidized films at $-196°C$. This may have been due simply to the small amount of material present.

B. Mode of Initiation of Autoxidation

A mode of initiation not considered above for autoxidizing cis-1,4-polyisoprene is from thermal cleavage of polyperoxides, formed by addition of peroxy radicals to double bonds. Such a reaction may occur in an inter- or intramolecular fashion.

$$RO_2\cdot + \; {>}C{=}C{<} \; \longrightarrow \; RO_2{-}C{-}C{<} \xrightarrow{\;O_2\;} RO_2{-}C{-}C{-}O_2\cdot \text{ etc.} \tag{23}$$

$$RO_2R' \longrightarrow RO\cdot + R'O\cdot$$

This case is mathematically equivalent to the first-order initiation reaction described above. On the other hand, studies of model olefins have shown that addition to the double bond [16] does compete with allylic H-abstraction [2] in all but specialized cases. Moreover, Hiatt and McCarrick [17] recently concluded that the rate of thermal decomposition of an allylic hydroperoxide, 2,3-dimethyl-3-hydroperoxybutene-1, could be accounted for entirely by an induced process involving polyperoxides formed by peroxy addition reactions. This result, while obtained from solutions of pure hydroperoxide in solvents and therefore under experimental conditions somewhat different than ours, suggests that the mixture of hydroperoxide and peroxide expected to form in autoxidizing *cis*-1,4-polyisoprene would behave as though only polyperoxide (unimolecular) initiation occurred. This result is consistent with the usefulness of Eq. (15) in describing the data (Fig. 4) (and contrary to our earlier conclusions), but these are severe approximations in applying our mathematical derivations to the system. This is particularly evident from the fact that maximum intensity of chemiluminescence appears in the polymer after 1 hr at 117°C and at a very high degree of oxidation and does not correspond, as we assumed in deriving Eqs. (14) and (15), to the time when the rate of oxygen absorption was at a maximum (Fig. 9).

The activation energies obtained from the temperature-jump experiments with autoxidized and singlet oxygenated polymer (21–25 kcal/mol) are much lower than the activation energies for simple peroxide cleavage, which are normally above 30 kcal/mol [12]. Although a number of terms comprise the observed activation energy (Eqs. (17) and (19)), most of these are expected to be small [18]* and appear in the equations in a compensating manner.

The activation energies from the low-temperature chemiluminescence are comparable to an activation energy of 25.8 kcal/mol calculated for the second-order decomposition of butenyl hydroperoxide [19].

Since singlet molecular oxygen was seen to react with *cis*-1,4-polyisoprene to produce a species where decomposition produced excited states, a quantum chain reaction is a possible pathway when the polymer autoxidizes, since these excited states can be quenched by triplet to produce additional singlet oxygen. Unfortunately, not enough information is available to allow a quantitative assessment of that point, but we can estimate that, if the excited state formed in Reaction 7 emits with an efficiency of 2.9×10^{-3} (the fluorescence yield of biacetyl [20]) and all other excited molecules produce singlet oxygen, then the maximum yield of the latter in *cis*-1,4-polyisoprene from this source after 1 hr at 117°C is about 10^{-7} gm/gm of polymer. This value is probably a lower limit, since biacetyl fluoresces more efficiently than most aliphatic

---

* Estimates for these terms are: $E_7 - E_6 = 3 \pm 2$ kcal/mol [8a], $E_6 < 3$ kcal/mol [18a], and $E_9 < 3$ kcal/mol for $^3O_2$ quenching (diffusion controlled [18b]).

ketones, but this source of singlet oxygen is probably not significant in comparison with Reaction 6, which has a quantum yield of at least 0.04 in one case [13].

To summarize, the chemiluminescence from autoxidizing cis-1,4-polyisoprene at high temperatures is complicated by multiple emitters, but the time and temperature dependence of the emission can be described fairly well by several equations based on simplified reaction schemes. Oxygen uptake measurements show that most of the chemiluminescence appears after the polymer has already undergone extensive oxidation. At low temperatures, partially autoxidized cis-1,4-polyisoprene and hydroperoxidized polyisoprene yield a similar value for the activation energy for initiation of ~ 25 kcal/mol, implicating an allylic hydroperoxide as a contributor to the initiation process.

ACKNOWLEDGMENTS

The authors thank Mr. Fleet Girod, Miss Michelle Birts, and Mr. Fred Moore for experimental assistance, Dr. Gerald Lundeen for helpful discussions, and Dr. Jim Hoyland for providing computer programs to carry out data analysis. This research was supported in part by the National Aeronautics and Space Administration under Contract No. NAS2-8195.

REFERENCES

1. *Rubber Chem. Technol.* **45**, 359–598 (1972); J. A. Howard, *ibid.* **47**, 976 (1974). [The former contains recent reviews of oxidation of 1,4-polyisoprene and additional references.]
2. J. A. Howard, "Free Radicals" (J. K. Kochi, ed.), Vol. II, pp. 3–62. Wiley, New York, 1973.
3. (a) R. F. Vassil'ev, "Progress in Reaction Kinetics" (G. Porter, ed.), Vol. 4, pp. 305–352. Permagon, Oxford, 1967. (b) V. Ya. Shlyapintokh, O. N. Karpukhin, L. M. Postnikov, V. F. Tsepalov, A. A. Vichutinskii, and I. V. Zakharov, "Chemiluminescence Techniques in Chemical Reactions." Consultants Bureau, New York, 1968; (c) R. F. Vassil'ev, *Russ. Chem. Rev.* **39**, 529 (1970); (d) R. E. Kellogg, *J. Amer. Chem. Soc.* **91**, 5433 (1969).
4. G. E. Ashby, *J. Polym. Sci.* **50**, 99 (1961).
5. (a) J. Stauff, H. Schmidkunz, and G. Hartmann, *Nature* **198**, 281 (1963); (b) M. P. Schard and C. A. Russell, *J. Appl. Polym. Sci.* **8**, 985, 997 (1964); (c) R. E. Barker, Jr., J. H. Daane, and P. M. Rentzepis, *J. Polym. Sci., Part A* **3**, 2033 (1964); (d) L. Reich and S. S. Stivala, *ibid.* **3**, 4299 (1965); (e) R. J. De Kock and P.A.H.M. Hol, *Int. Syn. Rubber Symp. Lect.* **4(2)**, 53 (1969); *Chem. Abstr.* **74**, 4425 (1971); (f) U. Isacsson and G. Wettermark, *Anal. Chim. Acta* **68**, 339 (1974); (g) T. V. Pokholok, O. N. Karpukhin, and V. Ya. Shlyapintokh, *J. Polym. Sci., Part A-1* **13**, 525 (1975).
6. G. D. Mendenhall, *Angew. Chem.* **89**, 220 (1977).
7. M. A. Golub, M. L. Rosenberg, and R. V. Gemmer, *Polym. Prepr. Amer. Chem. Soc., Div. Polym. Chem.* **17**, 699 (1976).
8. (a) G. Lundeen and R. Livingston, *Photochem. and Photobiol.* **4**, 1085 (1965); (b) A. V. Tobolsky, D. J. Metz, and R. B. Mesrobian, *J. Amer. Chem. Soc.* **72**, 1942 (1950).

9. D. R. Kearns, *Chem. Rev.* **71**, 395 (1971).
10. M. L. Kaplan and P. G. Kelleher, *Rubber Chem. Technol.* **45**, 423 (1972).
11. C. Walling and L. Heaton, *J. Amer. Chem. Soc.* **87**, 48 (1965).
12. R. Hiatt, "Organic Peroxides" (Daniel Swern, ed.), col. III, p. 27. Wiley (Interscience), New York, 1972.
13. J. A. Howard and K. U. Ingold, *J. Amer. Chem. Soc.* **90**, 1056 (1968).
14. M. A. Golub, M. S. Hsu and L. A. Wilson, *Rubber Chem. Technol.* **48**, 953 (1975).
15. S. W. Beavan and D. Phillips, *Eur. Polym. J.* **10**, 593 (1974); *Rubber Chem. Technol.* **48**, 692 (1975).
16. F. R. Mayo, *Accounts Chem. Res.* **1**, 193 (1968). [See also the references therein.]
17. R. Hiatt and R. McCarrick, *J. Amer. Chem. Soc.* **97**, 5234 (1975).
18. (a) J. A. Howard and J. E. Bennett, *Can. J. Chem.* **50**, 2374 (1972); (b) H. L. J. Backstrom and K. Sandros, *Acta Chem. Scand.* **12**, 823 (1958).
19. A. Chauvel, G. Clément, and J.-C. Balaceanu, *Bull. Soc. Chim. Fr.* 1774 (1962); *ibid.* 2025 (1963)
20. M. Almgren, *J. Mol. Photochem.* **4**, 327 (1972).

9

Raman Spectra of Biopolymers

BERNARD J. BULKIN

DEPARTMENT OF CHEMISTRY
POLYTECHNIC INSTITUTE OF NEW YORK
BROOKLYN, NEW YORK

I. INTRODUCTION

The Raman effect, discovered in 1929, has a history of applications to polymer problems, but this history contains a large fallow period. With the emergence of laser excited Raman spectroscopy in the early 1960s, there has been a succession of theoretical and experimental advances that have stimulated many new applications. In this paper a few of these applications to biopolymers are reviewed critically. In so doing, I have selected from the hundreds of papers published in this area a few that illustrate rather diverse problems. Thus, this is not meant as an exhaustive review of the literature.

II. EXPERIMENTAL TECHNIQUE

A. Background

The Raman effect is a light-scattering effect, and a second-order effect at that. It results from changes in polarizability in the course of a vibration. There has been considerable recent theoretical activity aimed at a better

121

understanding of the Raman effect.* Most of this work has yet to find application to other than small molecule problems. Indeed, much of this is just testing of the theory. Two developments are worth noting, however. First, Albrecht has greatly clarified the nature of resonance Raman scattering, the Raman effect occurring when the exciting radiation coincides with an electronic absorption [2]. This has considerable importance for studies of metal containing proteins, to be discussed in more detail later in this paper. Second, the derivation of a Raman tensor for an α helix by Fanconi et al. (3) was reported in 1969. The analysis is based essentially on a factor group–space group approach, similar to that used by others for crystals and nonhelical polymers [4].

Both the selection rules and the angular dependence of the Raman scattering can be predicted. To date, except for applications by Peticolas' research group, some of which will be presented below, there has been relatively little use made of these theoretical developments. In this regard, one might say that there has been a rapid jump over application of basic theory to simple systems to a more qualitative application of Raman spectroscopy to complex biopolymeric problems.

There have also been major recent advances in experimental apparatus. Much of the information in studying biopolymer problems via Raman spectroscopy is contained in intensities of the bands, ratios of intensities in different polarizations, temperature dependence of intensities, etc. In such measurements, total signal intensity is of paramount importance in determining accuracy of computed results. Great advances have been made in increasing Raman intensities. There are lasers with increased power, and when the sample can withstand the radiation intensity this is the most direct solution. There have been improvements in grating efficiency with the development of holographic gratings. In a monochromator with concave holographic gratings developed by the J–Y Corporation, the stray-light level is sufficiently low that we believe the monochromator can be used as a single monochromator. This means considerably higher throughput than one would achieve with a double or triple monochromator. For biopolymers this added luminosity is crucial. It can mean the difference between feasibility and unfeasibility of an experiment. There have been improvements in phtomultipliers that have improved sensitivity throughout the visible region, but particularly in the red region. Because some compounds can only be examined with red exciting radiation this again expands the range of problems to be examined by Raman spectroscopy. In our laboratory, we estimate that the combination of these three improvements has led to a hundredfold average improvement in accuracy of a Raman relative intensity measurement over

* See, for example, [1].

what we could attain with our first laser-excited Raman instrument in 1968.

Finally, an increasing number of workers have interfaced their Raman spectrometers to minicomputers. With signal averaging, excellent intensity measurements can be obtained even at lower laser powers.

To summarize the utilization of these advances, it can be stated that virtually all laboratories have high-power lasers available, and most have one of the new photomultipliers. However, to date there are only 15–30 computerized spectrometers in operation, and almost no laboratories have exploited the single monochromator.

With so many spectroscopic techniques available, an important question to ask is when does one turn to Raman spectroscopy to solve a problem? There has long been the idea that Raman spectra of aqueous solutions are an important alternative to infrared spectroscopy. This is certainly valid, as a consequence of the weak Raman scattering of water. But what about Raman spectroscopy viewed in the broader marketplace of structural tools—nuclear magnetic resonance, diffraction techniques, etc.?

Here I believe that a key feature is the applicability of Raman spectroscopy to the study of phases of widely varying fluidity or crystallinity. While x-ray diffraction provides relatively high resolution structural information in crystalline samples, and nuclear magnetic resonance (especially since the advent of high-frequency, multinuclear spectrometers) can provide such data in liquids, vibrational spectra give information that is of lower structural resolution and is less directly interpreted. This information is available, however, over the full range of phases from solid to liquid (gases too, but they are not relevant here). Thus, Raman spectra can be obtained on glasses, liquid crystals, plastic crystals, and other mesophases of intermediate order. As will be illustrated below, Raman provides an important link between structural information obtained in crystalline and solution phases. This link is of great importance for biopolymer problems.

Table I summarizes the positive aspects of application of the Raman effect to polymer problems. Only limited use has been made of isotopic substitution in Raman spectra of polymers, mainly in deuteration of exchangeable

TABLE I

Application of Raman Effect to
Polymer Problems—Positive Aspects

---

1. Spectra may be obtained in all phases
2. Aqueous solutions may be used
3. Sensitive to conformational change
4. Isotopic substitution affects spectra
5. Small sample size ($\sim$ 5-nl volume)

TABLE II

Application of Raman Effect to
Polymer Problems—Negative Aspects

1. Raman effect is weak
2. Fluorescence is $\sim \times 10^6$ as strong
3. Only moderate structural resolution
4. Conclusions often rely on indirect methods
5. Photochemistry may interfere

protons. However, there has been a spate of recent preparations of isotopically enriched polymers for nuclear magnetic resonance. It will be useful to obtain Raman spectra of these samples for structural checks. The small sample sizes are an obvious advantage for biopolymer problems, particularly when natural products are involved.

Table II summarizes the negative aspects of applying the Raman effect to polymers. Fluorescence has been a major interference. This can occasionally be eliminated by using a different exciting line, particularly one in the red region, from which follows the great importance of improved sensitivity in this region. An alternative to removing fluorescence in the data collection is to remove it in postprocessing. Because the fluorescence is invariably much broader than the Raman scattering (different time scales involved lead to different half widths) it can be filtered by computing the Fourier transform of the spectrum, picking the appropriate Fourier coefficients to eliminate low frequencies, and recomputing the inverse transform. This can also be used for smoothing of data.

We now turn to the application of Raman spectroscopy to biopolymer problems. Table III shows some problems, which will be illustrated below, where Raman spectroscopy has been successfully used in the solution. They are diverse and important. The study of porphyrins is carried out in a re-

TABLE III

Some Problems which can be
Successfully Treated by
Raman Spectroscopy

1. Peptide and protein conformation
2. Some nucleic acid conformation
3. S–S bonding changes
4. Porphyrin-related problems
5. Many T-dependent studies

TABLE IV

Some Problems where Raman Spectroscopy cannot now Provide Interpretable Data

1. Most biological multicomponent systems, unless resonance is involved
2. Enzyme-substrate and inhibitor interactions
3. Resonance problems involving quantitative intensities

sonance Raman mode. Table IV shows three areas where some attempts have been made to apply Raman spectroscopy, but where we believe the data cannot be interpreted with currently available theory or empirical justification. We now turn to several specific examples.

III. APPLICATIONS

A. Peptides and Proteins

Vibrational spectroscopy has long been used to study conformation in peptides and proteins. The so called amide (I) and (II) modes were widely used in infrared spectra, and these are also diagnostic in the Raman effect. In addition the amide (III) mode has been widely used in the Raman effect as a result of the pioneering work of Lord *et al.* [5] on the band.

Table V shows the amide (III) frequencies obtained on several simple biopolymers where structural information is available. It should be noted that the band tends to be weak for α-helical structures. This has occasionally led to incorrect assignments.*

TABLE V

Amide (III) Frequencies (cm^{-1}) in Polypeptides from Raman Measurements

Substance	α helical	β structure	Random coil or structure ionized
Polyglycine	1261	1234	
Poly-L-alanine	1261	1239	
Poly-L-glutamic acid			1248
Glucagon	1266	1232	1248
Poly-L-lysine	1311	1240	1243
Poly Ala-Gly	1271	1238	
Poly Ser-Gly	1264	1236	

* The problem is discussed in [6].

Poly-L-lysine provides a good illustration of the information obtainable from Raman spectra. Figure 1 shows spectra of the α-helical and β-sheet forms obtained by Peticolas *et al.* [7]. Frushour and Koenig [8] have also published on this and related polymers. Note first that good quality spectra can be obtained from reasonably dilute aqueous solutions. The amide (I) band of the α form is concealed by the strong water band, but appears clearly in the β sheet. Clear changes can also be noted in the amide (III) region and in the C_α–C stretching modes in the 945–990 cm^{-1} region. It has been shown from a study of these bands that in the transition from $\alpha \to \beta$ the polymer does not go through an intermediate structure. A quantitative plot of the temperature dependence of the amide (III) mode (relative intensity) is shown in Fig. 2, illustrating the ability of the Raman spectrum to measure the fraction of either conformation present. However, the idea works much better with synthetic biopolymers than with natural proteins, where a distribution of conformations is present. In such cases, the changes observed are likely to be small shifts in the frequency maximum of a broad band.

Lysozyme represents a well-studied protein that illustrates what can be learned. Lord [5] had shown that the frequency maximum of the amide (III)

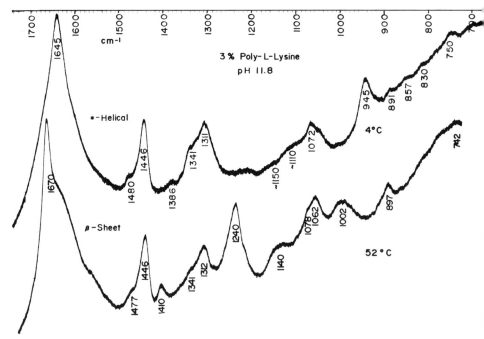

FIG. 1   Laser Raman spectra of (top) α-helical poly-L-lysine (3% H_2O, pH 11.8, $T = 4°C$) and (bottom) β-sheet poly-L-lysine (3% H_2O, pH 11.8, $T = 52°$). (From [7].)

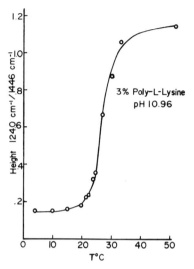

FIG. 2   Plot of the relative height of the 1240-cm⁻¹ band at pH 10.96 vs. temperature showing the α → β transition. (From [7].)

mode shifts on denaturation. More recently, Yu [9] utilized Raman spectroscopy to look at crystal vs. solution spectra for lysozyme. This is a case where x-ray work shows α, β, and random coil structures to be present. Yu found that the amide (III) band did not change shape between crystal and solution. He also noted that the amide (I) band showed just one peak, and was not as sensitive as the amide (III). Several changes were observed in the spectra, these being interpreted as changes in side-chain conformations. When spectra were taken of the lypholized powder of lysozyme, interesting changes were observed. The peak due to S–S stretching broadened, possibly indicating a greater distribution of dihedral angles about the S–S bond in this powder than is present in either the crystal or solution. There was also a small change in relative intensities in the amide (III) region, again indicative of conformational changes.

The denaturation of lysozyme can be carried out thermally in a reversible fashion, or it may be done irreversibly. The Raman spectra of Yu show that when the reversible denaturation is carried out there is not much conformational change. The S–S bonds clearly remain intact. In the irreversible denaturation, there is a decrease in the S–S stretching band intensity. This may not be due to a rupture of S–S bonds, necessarily, but could arise from a broader distribution of conformations. Figure 3 shows the results of semiempirical calculations of Boyd [10] on the dihedral angle of S–S bonds correlated with S–S stretching frequency and S–S overlap population. The Raman data used for the correlation are from Van Wart et al. [11].

For both polypeptide and proteins there has been some attempt to synthesize spectra from those of component amino acids and oligopeptides.

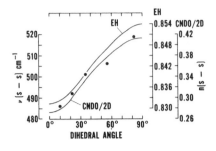

FIG. 3 Dependence of S–S stretching frequency v(S–S) and S–S overlap population n(S–S) on dihedral angle for disulfides. The Mulliken overlap populations are from deorthogonalized CNDO/2 and EH wave functions of the model compound H_2S_2 at a fixed S–S internuclear distance [11]. (From D. B. Boyd, *Int. J. Quantum Chem., Quantum Biol. Symp. No. 1*, 1974. Copyright © 1974 by John Wiley & Sons, Inc. Reprinted by permission of John Wiley & Sons, Inc.)

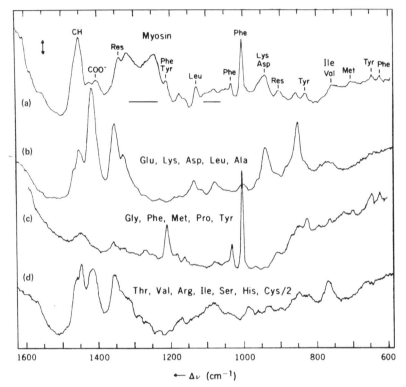

FIG. 4 Comparison of Raman spectra of aqueous solutions of (a) myosin and (b)–(d) its consituent amino acids. The relative concentrations of the amino acids correspond to those in myosin. The five most prevalent amino acids appear in (b); those in (c) are most common in the globular head portions of the myosin molecule. Horizontal lines indicate regions containing contributions from vibrations near the peptide bond. Abbreviations: Res, residue vibration; Gly, glycine; Thr, threonine; Ser, serine; Ile, isoleucine; His, histidine; Cys/2, cystine. Conditions were: (a) myosin, 35 mgpml in 0.6 M KCl, pH 7.0: (b)–(d) amino acids, pH 6.0, 3.0, and 11.0, respectively (to aid solution). Resolution, 5 cm α^1, scanning speed, 30 cm α^1/min; vertical arrow, 300 count/sec in (a) and 3000 count/sec in (b–d); laser power, 600 mW in (b–d); excitation, 5145 Å. (From E. B. Carew, I. M. Asher, and H. E. Stanley, *Science* **188**, 933 (1975). Copyright 1975 by the American Association for the Advancement of Science.)

This is an important activity that is just beginning to bear fruit. Perhaps the most interesting aspect of such syntheses are the spectroscopic regions where the data are not well reproduced. These regions are likely to be the most sensitive to long-range intra- and intermolecular effects.

Among the most ambitious attempts of this sort is the work by Carew et al. [12] on myosin. As in other proteins, the amide (III) region is shown to be characteristic of conformation for this case. Figure 4 shows the comparison of the Raman spectrum of myosin with that of its constituent amino acids. It highlights the ability to assign Raman peaks to particular types of residues, as well as the appearance of new bands such as the amide (III) modes in the 1230–1280 cm^{-1} region. Considerably more work needs to be done of this sort, particularly with computer-assisted summations of spectra.

B. Resonance Raman Spectroscopy

Conventionally, Raman spectra require concentrations on the order of 0.1 M or greater, which is of limited use for many biochemical problems. However, when the Raman exciting line falls within the envelope of an electronic absorption band, it becomes possible to see Raman spectra at concentrations of approximately 10^{-5} M. Qualitatively, these spectra arise only from the part of the molecule associated with the electronic transition. Thus, in the case of a protein like hemoglobin or a cytochrome containing a chromophore isolated (in a molecular orbital sense) from the bulk of the protein, we can see a selectively enhanced Raman spectrum.

Figure 5 shows the first spectra published for oxyhemoglobin and ferrocytochrome c by Spiro and Strekas [13]. Here the protein concentrations are on the order of 5×10^{-4} M. High quality (interpretable) Raman data are obtained from such experiments.

An interesting feature of these porphyrin spectra, in which symmetry is either D_{4h} or slightly reduced from D_{4h}, is that certain modes have large depolarization ratios. While in a nonresonance case the depolarization ratio varies from 0 (completely polarized) to 0.75 (depolarized), in this resonance case depolarization ratios much greated than 0.75 are seen. This was predicted from theory [14] and recognized by Spiro and Strekas when they observed it. The resonance Raman spectra of these proteins have since been correlated with structural studies from other techniques. This has been reviewed recently by Spiro [15].

While attention to date in resonance Raman spectroscopy of polymers has been focused on the proteins with localized chromophores, there are many other potentially interesting cases. It should be possible to examine structures of synthethic polymers where there is extended conjugation, for example. Because the resonance Raman effect involves essentially vibronic modes, it is sometimes useful for clarifying electronic absorption assignments.

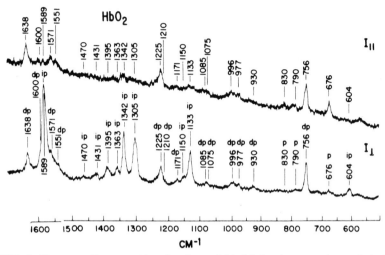

FIG. 5   Resonance Raman spectra of oxyhemoglobin (a) showing anomalous polarization. Both the direction and polarization vector of the incident laser radiation are perpendicular to the scattering direction. The scattered radiation is analyzed into components perpendicular $(I_\perp)$ and parallel $((I_{\|})$ to the incident polarization vector. The exciting wavelength was 568.2 nm. The slit width was 10 cm α^1. Concentration is 0.5 mM for HbO_2. (From [13].)

Further applications will emerge shortly as more experimental work is done on obtaining resonance Raman spectra in the ultraviolet. This is now being done successfully using argon or nitrogen lasers and frequency doublers. Once Raman spectra of benzene can be obtained in resonance, for example, one should be able to learn more about polymer problems such as the interactions between phenyl groups in PBLG.

C. Polynucleotides

Finally, there has been extensive work by a few groups on polynucleotide spectra, including much work on RNA and DNA structures and phase transitions. Much of this work has been reviewed recently by Thomas [16].

Nucleic acid–polynucleotide studies provide a clear example of the importance of Raman spectroscopy for providing solution structural information needed as a complement to x-ray diffraction. The structural problems for these systems often involve questions of exchangeable protons and helical forms that may interconvert in water. Raman spectra have thus been shown to be sensitive to primary structural changes such as tautomerization and ionization equilibria, as well as to secondary structural interactions, such as base pairing, base stacking, and backbone conformation. As in the case of peptide and protein spectra, it is proving possible to distinguish features

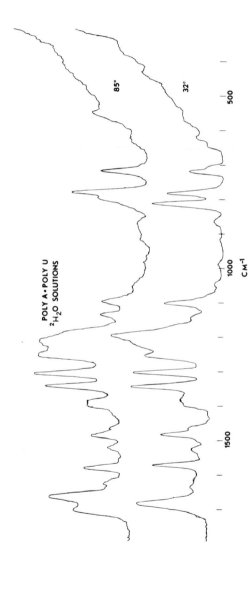

FIG. 6 Raman spectra of ^2H_2O solutions of poly(A)·poly(U) at 32°C and 85°C. Solute concentration is 35 mg/ml, spectral slit width 10 cm α^1, scan speed 50 cm α^1/min, and rise time 3 sec. (From L. Lafleur, J. Rice, and G. J. Thomas, Jr., *Biopolymers* **11** (1972). Copyright © 1972 by John Wiley & Sons, Inc. Reprinted by permission of John Wiley & Sons, Inc.)

associated with primary and secondary structures through synthetic spectra of oligonucleotides.

Thomas has carried out considerable work on the model polymers, e.g., poly(A), poly(U), poly(G), and poly(C) in such double helical forms as poly(A)·poly(U) and poly(G)·poly(C) [17]. A typical spectrum of poly(A)·poly(U) in ^2H_2O is shown in Fig. 6. The temperature dependence of this system illustrated here is quite striking. Since the change between these temperatures is dissociation from a double-stranded poly(A)·poly(U) to single-stranded poly(A) and poly(U), one can appreciate the sensitivity of Raman spectroscopy to such changes.

Raman spectra can be used to study reactions of these polymers. Thus there have been studies of the interaction of metal ions with bases and phosphate sites, nucleic acid–protein interactions, and alkylation of DNA and other polynucleotides. Thus far the alkylation studies have concentrated on spectra of the same polynucleotide or DNA before and after reaction [18]. However, since certain of the alkylating agents are highly colored, it may be of interest to examine these cases from the viewpoint of resonance Raman spectra of the alkylating agent as well.

Most recently, the work on polynucleotide–protein interactions elucidated by Raman spectra has led to the study of virus structure. Thus, for example, Turano et al. [19] were able to make use of many of the conclusions from model studies of polypeptides and polynucleotides to draw a substantial number of interesting conclusions from the Raman spectra of turnip yellow mosaic virus. The secondary structures of both the "coat" protein and the RNA were found to be different from those predicted. A number of other conclusions about the exposure of various amino acid residues to solvent could also be made. These and other conclusions are as good an indicator as any of the considerable progress in Raman spectra of biopolymers made in the last 15 years.

REFERENCES

1. A. C. Albrecht and M. C. Hutley, *J. Chem. Phys.* **55**, 4438 (1971).
2. A. C. Albrecht, *J. Chem. Phys.* **34**, 1476 (1961).
3. B. Fanconi, B. Thomlinson, L. A. Nafie, W. Small, and W. L. Peticolas, *J. Chem. Phys.* **51**, 3993 (1969).
4. R. Zbinden, "Infrared Spectroscopy of Polymers." Academic Press, New York, 1964.
5. R. C. Lord and N. T. Yu, *J. Mol. Biol.* **50**, 509 (1970).
6. M. C. Chen and R. C. Lord, *J. Amer. Chem. Soc.* **96**, 4750 (1974).
7. W. L. Peticolas, *Biochimie* **57**, 417 (1975).
8. B. G. Frushour and J. L. Koenig, *Biopolymers* **14**, 2115 (1975).
9. N. T. Yu, *Arch. Biochem. Biophys.* **156**, 469 (1973).
10. D. B. Boyd, *Int. J. Quantum Chem., Quantum Biol. Symp. No. 1*, pp. 13–19. Wiley (Interscience), New York, 1974.

11. H. E. Van Wart, A. Lewis, H. A. Scheraga, and F. D. Saeva, *Proc. Nat. Acad. Sci. U.S.A.* **70**, 2619 (1973).
12. E. B. Carew, I. M. Asher, and H. E. Stanley, *Science* **188**, 933 (1975).
13. T. G. Spiro and T. C. Strekas, *Proc. Natl. Acad. Sci. U.S.A.* **69**, 2622 (1972).
14. G. Placzek, *Handb. Radiol.* **2**, 209 (1934).
15. T. G. Spiro, *Biochim. Biophys. Acta* **416**, 169 (1975).
16. G. J. Thomas, Jr., "Structure and Conformation of Nucleic Acids and Protein–Nucleic Acid Interaction" (M. Sundraling and S. T. Rao, eds.), p. 253. University Park Press, Baltimore, 1975.
17. L. Lafleur, J. Rice, and G. J. Thomas, Jr., *Biopolymers* **11**, 2423, (1972).
18. S. Mansy and W. L. Peticolas, *Biochemistry* **15**, 2650 (1976).
19. T. A. Turano, K. A. Hartman, and G. J. Thomas, Jr., *J. Phys. Chem.* **80**, 1157 (1976).

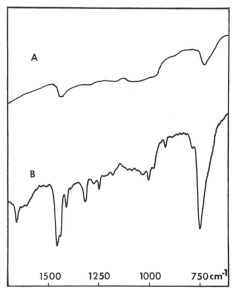

FIG. 1   Transmission infrared spectra in the range 500–1800 cm^{-1} of an Austin black filled polybutadiene (20 phr) dispersed in KBr: (A) conventional dispersive spectrometer (Perkin–Elmer 521); (B) FTIR (Digilab FTS-14). (From [1].)

14). We employed the identical sample dispersed in KBr to obtain these spectra. Note the marked increase in spectral detail of the FTIR spectrum.

Figure 2 shows the FTIR absorbance spectra of KBr pellets containing: (A), CB and Silene (SiO$_2$) EF (20 phr); (B), CB and GPF carbon black (20 phr); (C), CB and Austin black (20 phr); and (D), unfilled CB. In each case the spectra shown were operated on to remove the contribution from the KBr matrix and the sloping baseline [1]. As can be seen, the spectra are of excellent quality. Subtle band intensity and position changes are noted at 2915, 1450, 1433, 1310, and 738 cm^{-1}. These absorptions are characteristic of various C–H vibrations. The C=C stretching mode at 1655 cm^{-1} appears slightly affected. The doublet at 1450 cm^{-1} and 1433 cm^{-1} is presented in Fig. 3 for the unfilled elastomer and the GPF system. The intensity of the 1433-cm^{-1} band increases slightly with respect to the 1450-cm^{-1} absorption. These bands have been assigned to CH$_2$ bending vibrations of gauche and trans conformations, respectively, of freely rotatable C–C bonds [2]. This has also been clearly demonstrated in recent work on *trans*-1,4-polychloroprene [3]. The interaction between filler and polymer, therefore, apparently results in a loss of preferred conformation, conceivably because the polymer forms a strongly absorbed surface layer. It should be noted that we do not observe such changes in the Austin black system. Austin black is recognized

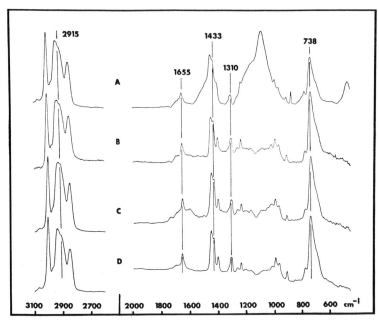

FIG. 2   Absorbance FTIR spectra in the range 500–3100 cm^{-1} of KBr pellets (after double substration) of: (A) CB and Silene EF (20 phr); (B) CB and GPF carbon black (20 phr); (C) CB and Austin black (20 phr); and (D) CB alone. (From [1].)

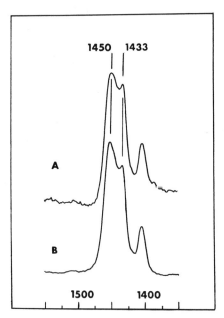

FIG. 3   Absorbance FTIR spectra in the range 1300–1600 cm^{-1} of KBr pellets (after double subtraction) of: (A) CB and GPF carbon black (20 phr); and (D) CD alone. (From [1].)

to be a weakly reinforcing carbon black whereas the GPF carbon black is considered to be strongly reinforcing. This evidence is by no means definitive and requires further work to account for such factors as filler particle size, a variable of obvious significance in the mechanism of interaction suggested above. It is intended to extend these studies and obtain the infrared spectrum of the so-called "carbon gel" (an intimate material composed of the elastomer and carbon black, which cannot be separated by the conventional solvent extraction procedures), which should accentuate these changes. Hopefully we can then gain information about how the carbon black interacts with the elastomer.

B. Coal and Related Materials

Coal is a complex, heterogeneous macromolecular system. Although a certain amount of structural information has been obtained by conventional infrared analysis, some formidable problems remain. The highly absorbing nature of the material makes the recording of high-quality infrared spectra difficult. In addition, such spectra are characterized by broad bands due to the overlapping absorptions of the numerous components.

FTIR has outstanding potential as a tool for coal characterization. High-quality spectra can be routinely obtained and demonstrate a marked superiority to spectra obtained on a dispersive instrument similar to that shown for carbon black filled polybutadiene discussed previously. High signal-to-noise ratio spectra obtained by multiscanning and signal averaging are of course essential to computer manipulation of data, such as spectral subtraction.

In this laboratory we have recently applied FTIR to coal characterization. Preliminary experiments have been mainly involved with the quantitative mineralogical analysis of the low temperature ash (LTA) of various coal samples through the application of the spectral subtraction method [4]. However, a more spectacular example of the type of new information that can be determined by FTIR is a method developed for the direct analysis of mineral matter in coal.

Figure 4a compares the spectrum of a coal with the spectrum of the same sample demineralized by acid treatment. The difference spectrum shown in the same figure was obtained by subtracting the bands characteristic of organic material in the latter spectrum from the spectrum of the former. The result should be a spectrum of the mineral matter in the coal, or more precisely the mineral matter that exists as a separate removable phase in the coal. This difference spectrum is compared to the spectrum of a low temperature ash of the same coal in Fig. 4b. They are not identical, but differ only in a few bands, noticeably the ones appearing at 1153 cm^{-1} and

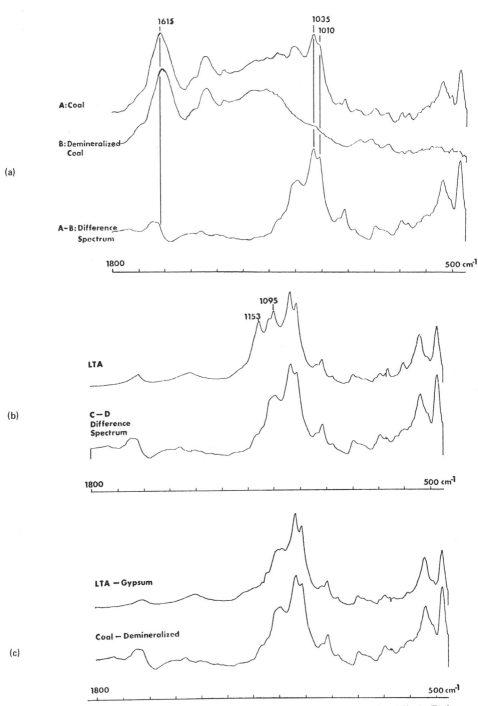

FIG. 4 (a) Absorbance FTIR spectra of: (A) coal (Illinois ≠ 6, Burning Star Mine); (B) demineralized coal; (A − B) difference spectrum characteristic of the mineral matter. (b) Comparison of absorbance FTIR spectra of the low temperature ash of coal (top) with (C − D)
the difference spectrum of Fig. 4a. (c) Comparison of difference spectrum obtained by subtracting the bands of gypsum from the spectrum of the low temperature ash with the difference
spectrum of Fig. 4a, obtained by subtracting the spectrum of the demineralized from that of
the parent coal.

1095 cm^{-1} characteristic of the hemihydrate of gypsum. If the spectrum of this latter mineral is subtracted from the LTA, then the final two difference spectra are almost identical, as shown in Fig. 4c. These results demonstrate the formation of gypsum in the LTA process, leading to anomalies in quantitative determinations based on the analysis of ash. In addition, this work strongly suggests that the spectral subtraction method used with such success in the characterization of polymers (see below) should be applicable to coal. We are currently studying various coal macerals and the products of liquefaction processess using this technique.

C. Degradation of Polyacrylonitrile

During the past two decades extensive research has been performed on polyacrylonitrile and its copolymers due to the fact that they are precursors of high-performance carbon fibers. At the present time our knowledge of the mechanisms operating in the oxidative process of carbon fiber formation from acrylic starting materials is limited. Detailed research is still required in areas such as: the role of oxygen introduced during oxidation and removed during carbonization; the cause of defects in the precursor acrylic fibers that

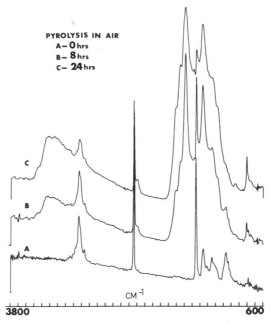

PYROLYSIS IN AIR
A— 0 hrs
B— 8 hrs
C— 24 hrs

FIG. 5   FTIR absorbance spectra of the range 600–3800 cm^{-1} of PAN at 200°C in air; (A) unpyrolysed; (B) pyrolysed for 8 hours; and (C) pyrolysed for 24 hours.

persist in the resultant carbon fibers and limit their strength; and the precise structure of the preoxidized product [5].

Recently, we have obtained FTIR spectra of degraded polyacrylonitrile that are far superior to those previously reported. Whereas spectra from conventional dispersive instruments tend to be very broad and unresolved at oxidation times in excess of ten hours at $\sim 200°C$, we obtain well-resolved spectra even after 24 hrs and are able to detect small but significant chemical changes with relative ease. Representative spectra of PAN degraded in air and vacuum for 24 hrs at 200°C in the range 700–3700 cm^{-1} are shown in Figs. 5 and 6. It is considered important to study concurrently the FTIR spectra of the pyrolyzed PAN in vacuum in order to better understand the complex spectrum obtained from air oxidation. Figure 7 shows the expanded spectrum in the range 2600–3000 cm^{-1} of the PAN degraded in vacuum for 16 hours. Note the excellent resolution and signal-to-noise ratio obtained even though the samples were highly colored. Of particular significance are the infrared bands at 1150, 1575, 1610, and those in the 3000–3500 cm^{-1} region observed in the vacuum-degraded PAN polymer. Many mechanisms have been proposed [5] and, although we are still in the initial stages of our

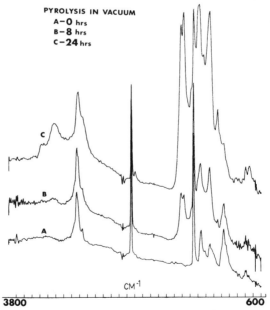

FIG. 6   FTIR absorbance spectra in the range 600–3800 cm^{-1} of PAN at 200°C in vacuum; (A) unpyrolysed; (B) pyrolysed for 8 hours; and (C) pyrolysed for 24 hours.

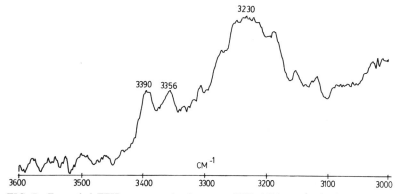

FIG. 7   Expanded FTIR spectrum in the range 3000–3600 cm^{-1} of PAN pyrolysed in vacuum at 200°C for 16 hours.

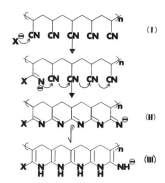

study, we are in a position to propose a tentative mechanism for the pyrolysis of PAN in vacuum that appears consistent with the infrared results. (See the mechanism shown here diagramatically.) The essential features of this mechanism are:

(1) Nucleophilic attack on a nitrile group (I) followed by cyclization of isotactic sequences along the chain (II).

(2) Rearrangement by tautomerism to give the structure (III). It is well known that —CH—C=N— structures tautomerize to —C=C—NH—, with the equilibrium strongly favoring the latter [6].

The FTIR data is consistent with the formation of (III). The band at 1610 cm^{-1} may be assigned predominantly to the C=C stretching vibration that appears at this frequency due to partial delocalization of electrons through the C–N bond. The 1575-cm^{-1} band is then assigned predominantly to the N–H bending vibration (a highly coupled mode), which together with the relatively broad band at 3230 cm^{-1} assigned to the N–H stretching vibration is strong evidence for the N–H structure. We have attempted to exchange the $>$N—H to $>$N—D using D_2O. Considering the impermeability of PAN we could only expect to deuterate the surface of the film but we did in fact observe a significant decrease in the 1575-cm^{-1} band compared to the 1610-cm^{-1} band. This supports the premise that the 1575-cm^{-1} band is due primarily to an N–H bending vibration. Obviously, FTIR studies of the degradation of the monodeutrated PAN, i.e., $+CH_2—CDCN+_n$, are crucial if we are to confirm this mechanism. We are currently preparing this polymer for these studies. The band at 1150 cm^{-1} is probably associated with the C–N stretching frequency while those at 3390 cm^{-1} and 3356 cm^{-1} are typical of primary amines. Protonation of the exocyclic imine (see (III)) would yield an amine consistent with these observations. The bands at 3188 cm^{-1} and 3150 cm^{-1} are most probably overtones and combination bands associated with the strong vibrations at 1610 cm^{-1} and 1575 cm^{-1}. Degradation in vacuum only gives one extra band in the C$\equiv$N stretching region at 2198 cm^{-1}, which could result from a β-imino nitrile formed from nucleophilic attack of a cyanide ion ($X^- = CN^-$ in (I)). It is also important to note that the C–H stretching frequency due to the tertiary α hydrogen at 2920 cm^{-1} disappears [7] upon vacuum degradation and that there is no infrared evidence for the presence of alkene or aromatic hydrogens.

It must be emphasized that the above mechanism, while simple and elegant, is only tentative and it is necessary to obtain infrared spectra of some crucial model compounds. Furthermore, we are currently studying the degradation of deutrated PAN, the base catalysized degradation of PAN, and also the degradation of polymethacrylonitrile. The results of these studies will have considerable significance for the above mechanism. Having accurately assigned the infrared bands occurring in the spectra obtained in vacuum we can then turn our attention to the more complex air-oxidized material. However, it is already evident that structure (III) with its highly activated methylene groups will readily oxidize to give quinone type structures. Hopefully these studies will assist in our understanding of the chemical processes occurring during degradation and lead to the prediction of superior carbon fibers.

III. SPECTRAL MANIPULATIONS

One of the more formidable problems in infrared spectroscopy is that the analytic bands of interest due to small concentrations of other structural units or irregularities, impurities, additives, oxidation or degradation prod-

ucts, etc., are often masked by the strong absorbance of the major component. Although not strictly limited to FTIR systems, the fact that the data is
stored digitally in the dedicated minicomputer allows one to readily perform
mathematical manipulations on the spectra. Consequently, it is feasible to
subtract the spectrum of a standard or unmodified polymer containing the
major backbone from that of the sample under study and accentuate the
differences between them. This type of procedure is more successful with
polymers containing large repeat units in which there is little vibrational
coupling. In the following sections we present examples of current research
where the ability to perform mathematical manipulation upon the experimentally derived spectra has significantly assisted in obtaining information
that was previously unattainable.

A. Polychloroprene

Polychloroprene is a polymer with a relatively large monomer repeating
unit that consequently exhibits little, if any, vibrational coupling between the
chemical units along the chain [8]. It is an outstanding example of a material
for which the digital absorbance subtraction method may be successfully
employed to obtain significant information that was previously unattainable.
Thus we are able to separate the infrared bands due to the crystalline phase
of the material from those of the amorphous [3] and, in addition, accentuate
and identify infrared bands due to the presence of structural irregularities
present in the otherwise predominantly trans-1,4 configuration [9].

Initially, we were interested in using the subtraction method to obtain a
"purified" spectrum of the crystalline phase of the material [3]. This technique considerably simplifies the assignment of the individual infrared bands
arising from the crystalline and amorphous components of the spectrum
and is important for subsequent theoretical normal coordinate calculations
[8]. The procedure is illustrated in Fig. 8, where we have subtracted the
amorphous contribution of a chloroprene polymer polymerized at $-150°C$
[10] from that of the semicrystalline material. This polymer consists of
essentially pure trans-1,4 units with only approximately 2% as head-to-head
placements and was not available when we published our previous results.
The amorphous spectrum was recorded at 80°C in a heated vacuum cell and
the amorphous band at 602 cm^{-1} was employed to determine the correct
subtraction parameter. A slight and unavoidable oxidation of the heated
polymer is evident from the bands at 1734 cm^{-1} and 1802 cm^{-1} but causes
an insignificant effect to the resulting difference spectrum because of the
relatively small concentration of the amorphous component of this polymer.

Previous studies on chloroprene polymers polymerized at temperatures
in the range $-20°C$ to $+40°C$ revealed that specific crystalline bands
(obtained by the subtraction technique described above) were sensitive to the

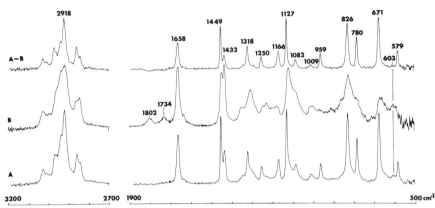

FIG. 8   FTIR spectra in the range 500–3200 cm^{-1} of chloroprene polymerized at $-150°C$; (A) spectrum recorded at room temperature; (B) spectrum recorded at 80°C; and (A − B) difference spectrum obtained by subtracting (B) from (A).

number of structural irregularities present in the polymer. It is well known that as the polymerization temperature is increased, the concentration of head-to-head trans-1,4; cis-1,4; 1,2 and 3,4 units also increases [11]. From these studies, and those of a copolymer of chloroprene and 2,3-dichlorobutadiene, it was concluded that specific structural irregularities are incorporated into the crystalline lattice, causing a perturbation of the vibrational force field that results in specific frequency shifts. It is instructive to include the data we have recently obtained on the $-150°C$ polymer (see Table I). The most sensitive bands are those at 959, 1009, 1250, and 1318 cm^{-1} ($-150°C$ polymer). These are all highly coupled normal modes [8] involving the C–C stretch and various CH_2 bending vibrations. Originally, based on previous infrared and NMR results, we concluded that it was probable that the cis-1,4 unit was incorporated into the crystalline phase, causing the observed frequently shifts. However, recent ^{13}C NMR results [12,13] suggest that the concentration of cis-1,4 units in the chloroprene polymers has been considerably overestimated (e.g., 5.2% by ^{13}C NMR compared to 13% by infrared for the $+40°C$ polymer). It is now considered more likely that it is the inverted (head-to-head) trans-1,4 units and not the cis-1,4 units that are incorporated into the crystalline lattice. An examination of the molecular models of these two structural irregularities supports this premise. The presence of an inverted trans-1,4 unit in an otherwise head-to-tail trans-1,4 polymer causes a relatively minor effect on the conformation of the chain and could be slowly incorporated into the lattice ([12], and see Fig. 4). Conversely, the cis-1,4 unit causes a major perturbation of the chain conformation and is likely to be excluded for the crystalline lattice. It is also significant

TABLE I

Crystalline Vibration Frequencies (cm^{-1}) *of*
trans-1,4-Polychloroprene as Function of Polymerization
Temperature

$-150°C$	$-20°C$	$0°C$	$40°C$
1658	1660	1660	1660
1449	1449	1448	1447
1318	1318	1316	1313
1250	1250	1252	1254
1166	1167	1167	1167
1127	1127	1127	1127
1082	1083	1083	1083
1009	1007	1005	1004
959	958	954	953
826	826	826	826
780	780	779	778
671	671	671	671
579	577	576.5	576

that the copolymer of chloroprene and 2,3-dichlorobutadiene polymerized at $-20°C$, which effectively introduces 2,3-dichloro-2-butenylene structural irregularities into the polymer chain, shows the same frequency shifts and indicates that this structural unit is incorporated into the crystalline lattice. The presence of a 2,3-dichloro-2-butenylene unit in an otherwise trans-1,4 polychloroprene chain is somewhat analogous to the incorporation of an inverted trans-1,4 unit.

Additional information may be obtained by studying the amorphous spectra of the various polychloroprenes prepared at different polymerization temperatures. By digitally subtracting the amorphous spectrum (acquired at $70°C$ in vacuum) of a more structurally regular polychloroprene from that of a polymer containing a higher concentration of structural irregularities, we can accentuate the infrared bands due to these structural irregularities. Figure 9 illustrates this procedure. Spectra (A) and (B) are those obtained at $70°C$ from the polychloroprene samples polymerized at $-20°C$ and $40°C$ respectively. The scale expanded difference spectrum $(B - A)$ was obtained using the bands attributable to trans-1,4 units at 1660, 1305, 1118, and 825 cm^{-1} to determine the correct subtraction parameter. In terms of experimental technique, it should be emphasized that to obtain high signal-to-noise difference spectra it is essential that the films employed should be of comparable thickness (<0.5 mil) and be within the absorbance range where the Beer–Lambert law is obeyed. The infrared bands observed in the difference spectrum (see Table II) are attributable to the cis-1,4; 1,2 and 3,4

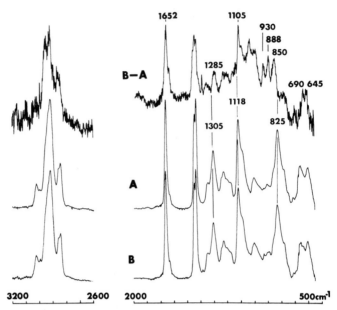

FIG. 9 FTIR spectra at 70°C in vacuum in the range 500–3200 cm^{-1}: (A) polychloroprene polymerized at −20°C; (B) polychloroprene polymerized at +40°C; (B − A) difference spectrum obtained by subtracting (B) from (A). (From [9].)

TABLE II

Major Infrared Bands Observed in the Difference Spectrum of the (+40)–(−20)°C Polychloroprene at 70°C

cm^{-1}	Tentative assignments	
1652 (vvs)	cis-1,4	C=C stretch
1635 (m)	1,2:3,4	C=C stretch
1440 (vs)	cis-1,4	CH$_2$ deformation
1430 (vs)	cis-1,4	CH$_2$ deformation
1360 (w)		
1285 (ms)	cis-1,4	
1225 (ms–sh)	cis-1,4	
1205 (ms)	cis-1,4	
1130 (w)		
1105 (s)	cis-1,4	
1090 (ms–sh)	cis-1,4	
1030 (vs)	cis-1,4; 1,2 (?)	
985 (vs)	1,2	*trans*-CH Wag
930 (s)	1,2	CH$_2$ wag
888 (s)	3,4	CH$_2$ wag
850 (s)	cis-1,4	
690 (ms)	cis-1,4	cis-CH wag
645 (ms)	cis-1,4	

structural units and represent a difference of approximately 4,1, and 1%
respectively [13]. By examining the infrared spectra of the model compounds
cis- and trans-4-chlorooctene, 3-chlorobutene-1, 1-chlorobutene-2, and 2-
chlorobutene-1 we can tentatively assign the major bands occurring in the
difference spectrum. Of particular interest is the fact that we have been able
to resolve the $C{=}C$ stretching vibration of the cis-1,4 unit at 1652 cm^{-1} from
that of the very strongly absorbing 1660 cm^{-1} $C{=}C$ stretch of the predominant trans-1,4 unit (see Fig. 10). The major bands due to the cis-1,4 unit
(i.e., 1652, 1440, 1430, 1285, 1225, 1205, 1105, 1090, 1030 850, 690, and 645
cm^{-1}) observed in the difference spectrum are consistent with the infrared
spectrum reported previously for the predominantly cis-1,4-polychloroprene
[14]. The 1,2-unit is readily identified by the well-isolated bands at 985 and
930 cm^{-1} while the 3,4-unit has a useful band at 888 cm^{-1}. There are no
isolated bands that can be unambiguously assigned to the isomerized 1,2-unit
[12] although a weak band occuring at 785 cm^{-1} is consistent with a medium
intensity band occurring in the model compound. Finally, it is feasible to
use the above method to obtain semiquantitative data on the concentration

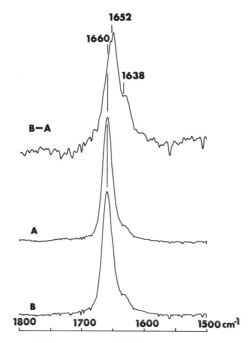

FIG. 10   Expanded FTIR spectra at 70°C in vacuum in the range 1500–1800 cm^{-1}:
(A) polychloroprene polymerized at −20°C; (B) polychloroprene polymerized at +40°C; and
(B − A) difference spectrum obtained by subtraction (B) from (A). (From [10].)

of structural irregularities in polychloroprene. If an amorphous spectrum of the structurally regular -150°C polymer is employed as a standard and subtracted from that of the polychloroprene under investigation, the concentration of structural irregularities could be calculated from a knowledge of the subtraction parameter. This would, however, necessitate accurately weighing the sample contained in the infrared beam.

Using polychloroprene as an example, subtraction techniques can be employed to gain further useful information:

(a) Detection and identification of additivites. If the infrared spectrum of a standard pure polymer is obtained and stored in the computer it can be used to remove the base polymer spectrum from the spectrum of an unknown polychloroprene sample. It thus becomes feasible to detect and identify additives such as antioxidants, vulcanizing agents, fillers, etc.

(b) End-group analysis. The detection and identification of polymer end groups can be of considerable importance. If a standard polymer of very high molecular weight with established end groups can be prepared, it can be used to subtract the base polymer infrared spectrum from that of an unknown sample of lower molecular weight. As the concentration of end groups is greater in the latter polymer sample they will be accentuated on subtraction of the standard.

(c) Oxidation and degradation. Infrared analysis has been extensively employed in the past to study the oxidation and degradation of polymeric materials. However, these studies have usually been performed on materials in rather advanced stages of oxidation or degradation. By subtracting the pure polymer from that of the oxidized or degraded sample we can now gain information concerning the initial stages of oxidation or degradation [15]. An example is shown in Fig. 11.

The unoxidized amorphous absorbance spectrum obtained at 62°C of the polychloroprene polymerized at 40°C is designated (A). After slight oxidation at 62°C while in the spectrometer the spectrum (B) is obtained. A 1:1 subtraction of the spectra (A) and (B) yields the difference spectrum (B − A). Those bands that appear positive are associated with the oxidized product while those appearing negative, i.e., 1660, 1305, 1118, and 825 cm^{-1}, are due to the loss of trans-1,4 structural units. The bands at 1802 and 1734 have previously been observed [16] and are assigned to an acid chloride and and aldehyde, respectively. However, the presence of the 965- and 920-cm^{-1} bands, which are probably associated with vinyl (CH_2=CH—) groups, have not been previously observed and their presence could have a significant effect on the proposed mechanism of oxidation. We are currently studying the spectra obtained as a function of the degree of oxidation.

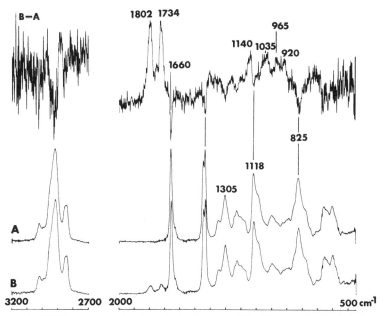

FIG. 11   FTIR spectra at 62°C in air in the range 500–3200 cm^{-1} of polychloroprene polymerized at 40°C: (A) unoxidized; (B) slightly oxidized; and (B − A) difference spectrum obtained by subtracting (A) from (B).

B. Polyvinylidene Chloride

Thermal analysis (DTA and DSC) has previously been used to study the melting behavior of polyvinylidene chloride (PVDC) single crystals [17]. The thermograms were found to contain multiple peaks in the melting region of the polymer that were sensitive to relatively small changes in annealing and crystallization conditions. These results were tentatively interpreted as concurrent degradation and thickening by annealing of the PVDC.

Initially, we employed FTIR spectroscopy to determine whether, in fact, significant degradation had occurred under the same annealing conditions as those used for thermal analysis. We were unable to detect the presence of any unsaturation or carbonyl formation in the 1550–1800 cm^{-1} region of the spectrum even after the longest annealing times. This led to the conclusion that the occurrence of degradation could not explain the thermal analysis results. We then turned our attention to the possibility that PVDC might exist in more than one polymorphic form similar to those observed in polyvinylidene fluoride [18,19].

The FTIR spectrum of PVDC single crystals is shown in Fig. 12. The vibrational spectra of PVDC has previously been studied [20–23]. It was

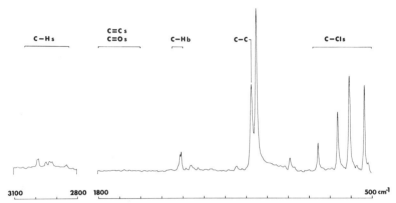

FIG. 12   FTIR spectrum of polyvinylidene chloride single crystals in the range 500–3100 cm⁻¹.

initially surmised that if PVDC existed in polymorphic forms, the C–Cl stretching frequencies should be sensitive to conformational changes. However, the most observable differences were detected in the C—H stretching region of the spectrum. This is a relatively weakly absorbing region of the spectrum for this polymer (see Fig. 12) and is not generally considered to be very sensitive to conformational changes. Taking advantage of the signal averaging capabilities of FTIR, which results in an excellent signal-to-noise ratio, we were able to accentuate this region of the spectrum.

Figure 13 shows the C–H stretching region (2700–3200 cm⁻¹) of the "as polymerized" and single crystals of PVDC prepared at 90°C. Four peaks were observed [21] for PVDC in this region at 2990, 2948, 2930, and 2850 cm⁻¹. These results were taken as evidence for a TGTG′ chain conformation that has symmetry isomorphous to the point group C_s and predicts four infrared active C–H stretching vibrations. However, the spectrum of the PVDC single crystals shows the presence of six distinct bands at 2849, 2917, 2930, 2948, 2981, and 2987 cm⁻¹. Close examination of the "as polymerized" PVDC also indicates the presence of a shoulder at approximately 2917 cm⁻¹. These results cast doubt on the previous assignments and interpretation.

In order to confirm the PVDC single crystal results we decided to prepare single crystals at different crystallization (T_c) and solution (T_s) temperatures to ascertain whether the intensities of the observed bands were sensitive to crystallization conditions. Figure 14 shows the C–H stretching region of two PVDC single crystal samples prepared at a T_c of (A) 85°C and (B) 90°C. The latter sample only reveals four infrared bands at 2930, 2948, 2981, and 2987 cm⁻¹, while the former exhibit major bands at 2917 cm⁻¹ and 2849 cm⁻¹ together with the bands at 2987, 2981, 2948, and 2930 cm⁻¹. It is evident that

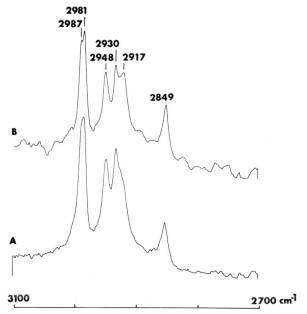

FIG. 13   FTIR spectra of PVDC in the range 2700–3100 cm^{-1}: (A) "as polymerized"; and (B) single crystals grown at 90°C.

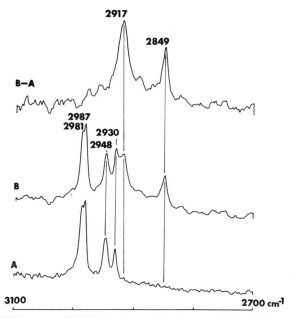

FIG. 14   FTIR spectra of PVDC single crystals in the range 2700–3100 cm^{-1}: (A) crystals grown at 85°C; (B) crystals grown at 90°C; and (B − A) difference spectrum obtained by subtracting (A) from (B).

the infrared spectra of the PVDC single crystals (Fig. 13) is a composite of two different spectral contributions. Furthermore, using digital absorbance subtraction it is feasible to obtain a "purified" spectrum (B − A) of the second component of the spectrum of PVDC that displays only two major bands at 2917 cm^{-1} and 2849 cm^{-1}.

Before we proceed to the other regions of the spectrum it is instructive to review the possible chain conformations of PVDC and to consider these in light of the above results.

The interpretation of the x-ray data obtained from PVDC is not unambiguous. The chain axis identity period is approximately 4.7 Å indicating two monomer units per repeat, which effectively rules out the planar zigzag conformation. A 2_1 helix was proposed [24] that has a chain symmetry isomorphous to the point group C_{2v} and predicts three infrared active C–H stretching frequencies—two σ bands of fairly strong intensity and one weaker π band [21]. A TGTG′ conformation [25] isomorphous to the point group C_s predicts four infrared active C–H bands (two σ bands of relatively strong intensity and two weaker π bands). It is also possible that the chain conformation is in an S_n helix (e.g., 10_3, 8_3, etc.), which has been suggested for polyisobutylene. We would expect to observe two C–H stretching frequencies for such a chain conformation because in most helical polymers the A and E modes are superimposed. Finally, we must also consider the possibility of bands associated with amorphous or random conformations.

From our experimental data, we conclude that the conformation of the chain in the crystals that exhibit only four bands (Fig. 14A) are consistant with the TGTG′ model. We tentatively assign the bands as follows:

$$2987 \; v_a(CH_2)_o \; A'$$
$$2981 \; v_a(CH_2)_i \; A''$$
$$2948 \; v_s(CH_2)_o \; A'$$
$$2930 \; v_s(CH_2)_i \; A''$$

The bands appearing at 2849 cm^{-1} and 2917 cm^{-1} cannot be unambiguously assigned at this stage. It is difficult but not impossible to believe that these bands are due to an amorphous component. However, we do not observe any contribution from these bands in the spectrum of PVDC crystals grown at 85°C (Fig. 14A), which would imply an almost perfect crystalline material if these indeed were amorphous bands. A further possibility is that we are observing bands due to a second crystalline form.

Figure 15 shows the 1200–1600 cm^{-1} region of the spectrum for PVDC crystals. The spectrum of the crystals prepared at 85°C indicates only two CH$_2$ bending vibrations at 1403 and 1410 cm^{-1}. This is consistent with the TGTG′ model with assignments of the 1403 cm^{-1} to the $\delta(CH_2)_o$ and

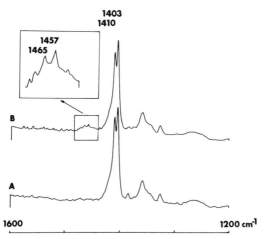

FIG. 15   FTIR spectra of PVDC single crystals in the range 1200–1600 cm^{-1}: (A) crystals grown at 85°C; (B) crystals grown at 90°C; and insert, expansion of 1440–1480 cm^{-1} region.

1410 cm^{-1} to the $\delta(CH_2)_i$ modes. A singlet at 1407 cm^{-1} had previously been observed [21] and assigned to the $\delta(CH_2)_i$ A′ mode. There is no evidence for a band at 1460 cm^{-1} tentatively assigned by the same author to an amorphous CH_2 bending mode.

The crystals prepared at 90°C shows four CH_2 bending modes (Fig. 15B). By digital absorbance subtraction, employing the same criteria as that used for the C–H stretching region, we have been able to eliminate the strong 1403/1410 doublet, leaving only the two bands at 1457 cm^{-1} and 1465 cm^{-1}. These latter bands appear to be solely associated with the 2849- and 2917-cm^{-1} bands occuring in the C–H region of the spectrum.

Obviously it would be advantageous to extend the above procedure to the C–C and C–Cl stretching region of the spectrum employing the identical substraction parameter used for the C–H stretching and CH_2 bending regions. However, this points out a major limitation of the subtraction procedure. Reasonable absorbance subtraction can only be carried out in the Beer's law range and the bands associated with the C–Cl stretching frequency are considerably stronger than those of the CH_2 stretching and bending modes (see Fig. 12). As a consequence it is necessary to employ a smaller sample size for subtraction in the C–Cl region. Preliminary results are not definitive. It appears that there are minor frequency shifts in the four intense C–Cl bands associated with the TGTG′ conformation in the presence of significant amounts of the other component. This is reminiscent of the frequency shifts observed for crystalline *trans*-1,4-polychloroprene containing structural irregularities incorporated in the lattice [26]. It is feasible that a mixed crystalline phase of PVDC could perturbate the vibrational force field, leading to subtle frequency shifts of the C–Cl vibrational modes. In any case, subtraction in this region is not "perfect" and leads to derivative type bands.

Finally, we have found that PVDC crystals that are essentially composed of pure TGTG′ conformation slowly transform at room temperature (see Fig. 16). This is evidence for the assumption that material associated with the 2849- and 2917-cm^{-1} bands are not due to an amorphous conformation. It is difficult to believe that on standing at room temperature the amorphous contribution of a crystalline polymer would increase.

In conclusion, we suggest, contrary to the generally held belief, that the C–H stretching region is rich in spectral detail and is quite sensitive to conformational changes. We are currently extending these studies to include other vinylidene polymers and will report our results in the near future.

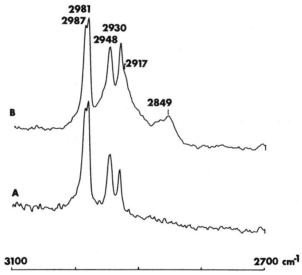

FIG. 16   FTIR spectra of PVDC single crystals in the range 2700–3100 cm^{-1}: (A) original crystals; (B) after 11 days at room temperature.

C. Polymer Blends—Polyvinylidene Fluoride–Polymethyl Methacrylate

In recent years there has been considerable interest in polymer blends and, in particular, the subject of mutual compatibility of the component polymers [27]. One of the more fascinating polymer pairs that has been reported to be compatible over a range of compositions in the solid state is the polyvinylidene fluoride (PVDF)–atactic polymethyl methacrylate (PMMA) system [28–30]. This is somewhat surprising when one considers the fact that PVDF is a highly crystallizable polymer that exhibits polymorphism [18,19]. A possible explanation for the apparent compatibility of the two polymers is that some specific interaction (e.g., complex formation) occurs between the individual PVDF and PMMA chains. Recent studies [31] have been reported on the ^{13}C spin-lattice relaxation (T_1) of N,N–dimethylformamide (DMF) solutions of the pure PVDF and PMMA and a mixture of the two polymers. They observed no significant differences in the T_1 values, which suggests that complex formation does not occur between PVDF and PMMA in solution. However, DMF may well interact with the polymers and prevent complex formation between the PVDF and PMMA.

Thin films of PVDF, PMMA, and mixtures of the two polymers were prepared from solutions of acetone just below its boiling point by casting onto potassium bromide (KBr) windows. The solvent was then completely removed by vacuum desiccation at room temperature. It was essential for subsequent spectral manipulation to ensure that films were sufficiently thin to be within the absorbance range where the Beer–Lambert law is obeyed [9,15].

It was postulated that FTIR studies of the PVDF/PMMA blend system should be very revealing based on the following, somewhat simplistic, argument [31a]. If the two polymers are completely incompatible, which implies phase separation, it should be possible to synthesize the infrared spectrum of the blend by the weighted absorbance coaddition of the spectra of the two pure components. In other words, neither of the two individual polymers recognizes, in infrared spectral terms, the existence of the other polymer in the blend. On the other hand, if indeed the polymers are completely compatible it implies that there is only one phase and that the individual polymer chains of both polymers are at least randomly dispersed. In fact, it is more probable that a distinct chemical interaction exists between the chains of the one polymer with that of the other. In this case, there should be considerable differences between the experimental infrared spectrum obtained from blend compared to that synthesized from the coaddition of the absorbance spectra of the pure components. Not only would these differences be derived from chemical interactions resulting in band shifts and broadening but also from changes in conformation.

Figure 17 shows the FTIR spectra in the range 700–2000 cm^{-1} of pure PMMA (D), pure PVDF (C), and a blend of PVDF/PMMA containing 75:25 parts by weight (B). This blend is considered to be incompatible by virtue of its composition and opacity. The spectrum designated (A) was obtained by coadding the absorbance spectra of the pure components (C) and (D) using appropriate weighting factors. It is apparent that the synthesized spectrum (A) is almost identical to that of the experimentally obtained spectrum of the blend (B).

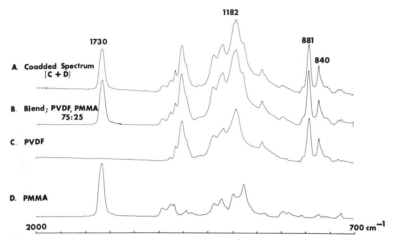

FIG. 17   FTIR spectra in the range 700–2000 cm⁻¹: (A) coadded (synthesized) spectrum of PVDF and PMMA; (B) experimental blend spectrum of PVDF and PMMA (72:25 parts by weight); (C) spectrum of pure PVDF; and (D) spectrum of pure PMMA.

Figure 18 shows the results of an identical experiment using a PVDF/PMMA blend composition of 39:61 parts by weight (equivalent to a 1:1 mole ratio) respectively. This blend is considered compatible by virtue of its composition and clarity. We were unable to approximate the experimental spectrum of the blend by absorbance coaddition of the two pure polymer

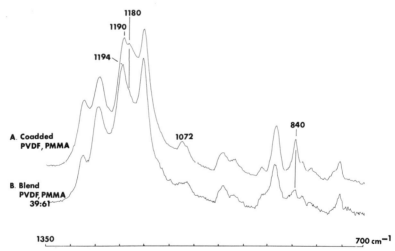

FIG. 18   FTIR spectra in the range 700–1350 cm⁻¹: (A) coadded (synthesized) spectrum of PVDF and PMMA; and (B) experimental blend spectrum of PVDF and PMMA (39:61 parts by weight). (From [31a].)

spectra. In this case we have expanded the 700–1350 cm^{-1} region of the spectrum to accentuate the most prominent differences. The synthesized spectrum (A) was obtained by coadding the spectra of PVDF and PMMA in such proportions as to yield an "incomparable" composite spectrum equivalent in terms of relative intensities to a 39:61 PVDF:PMMA blend. The weighting factors were derived from those employed for the incompatible 75:25 PVDF:PMMA blend discussed in the preceding paragraph. It is apparent that the experimental compatible blend spectrum (B) is significantly different from that of the synthesized spectrum (A). In particular, bands at 1180, 1072, and 840 cm^{-1} are not matched. They are considerably stronger in the synthesized spectrum (A) compared to that of the blend (B). By adjusting the weighting parameters for the coaddition of the pure PVDF and PMMA spectra we were unable to obtain a better matching of the experimental blend spectrum. We consider these spectral changes evidence for chemical interactions or conformational changes in the compatible blend.

In order to determine such conformational and interactional differences it is instructive to employ digital absorbance subtraction techniques. Figure 19 shows the infrared spectra of the incompatible 72:25 PVDF/PMMA, pure PMMA, and the difference spectrum obtained by subtracting the PMMA spectrum from that of the blend. We used the C=O stretching frequency of the PMMA at 1730 cm^{-1} to determine the correct subtraction parameter. It can be seen that complete subtraction of this band is not possible due

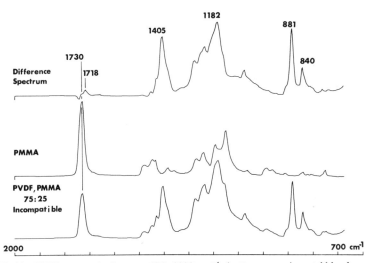

FIG. 19   FTIR spectra in the range 700–2000 cm^{-1}: bottom, experimental blend spectrum of PVDF and PMMA (75:25 parts by weight); middle, spectrum of pure PMMA; and top, difference spectrum blend – PMMA. (From [31a].)

primarily to band broadening and shifting of the C═O stretching frequency in the blend. A component at approximately 1718 cm^{-1} is evident and thus is strong evidence for some small but distinct interaction involving the C═O group. However, the difference spectrum is basically the same as that observed for pure PVDF (see below). This suggests that the majority of the PVDF in this blend is unaffected by the presence of the PMMA.

In contrast, the difference spectrum of the compatible 25:75 PVDF/PMMA blend obtained under the same conditions as described in the previous paragraph, indicates that distinct interactions have occurred involving the C═O group. This is illustrated in Fig. 20. In addition it is apparent that the difference spectrum differs considerably from that of pure PVDF, as shown in Fig. 21. Note particularly the apparent absence of the 840- and 1073-cm^{-1} bands and the shifting of the band at approximately 1200 cm^{-1}. It is well known that PVDF exists in at least three polymorphic forms [18,19] and that the infrared spectra of these three forms are markedly different. From previous studies [19] it was suggested that casting films of PVDF from hot acetone solutions results in the formation predominantly of form (II), a TGTG' conformation in the crystal. However, we found from our spectra of the pure PVDF and the difference spectrum obtained from the incompatible blend (both cast from hot acetone solutions) that the PVDF was predominantly of form (I), a planar zigzag conformation. This is consistent with the observations [18] that predominantly form (I) is obtained by casting

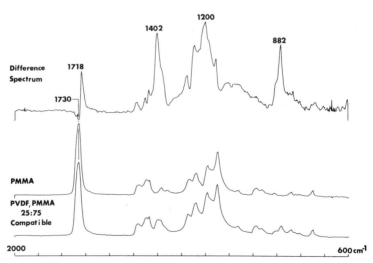

FIG. 20   FTIR spectra in the range 700–2000 cm^{-1}: bottom, experimental blend spectrum of PVDF and PMMA (25:75 parts by weight); middle, spectrum of pure PMMA; and top, difference spectrum blend − PMMA. (From [31a].)

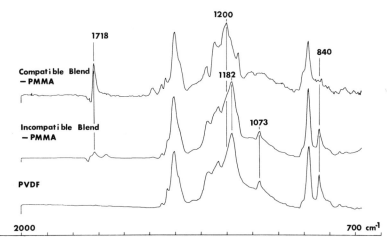

FIG. 21   FTIR spectra in the range 700–2000 cm^{-1}: bottom, spectrum of pure PVDF; middle, difference spectrum obtained by subtracting pure PMMA from PVDF/PMMA (75:25 parts by weight); and top, difference spectrum obtained by subtracting pure PMMA from PVDF/PMMA (25:75 parts by weight). (From [31a].)

from hot MEK. The difference spectrum of the compatible blend, however, does not appear to be consistent with either form. Nevertheless, the most important point is that if we assume an interaction between the PVDF and PMMA we would expect to observe conformational changes in the PVDF compared to that present in the crystal. A conformation for which optimum interaction might be attained between the different polymer chains is not unreasonable. Obviously, with an atactic PMMA polymer it is not feasible that a complete register of the interacting units could be attained.

Hence, we conclude that FTIR studies of PVDF/PMMA blends shows strong evidence of chemical interaction between the polymer chains of the two components that could be responsible for the apparent compatibility of of this system in a specific range of compositions.

D. Polyethylene Single Crystals in Suspension

In broad outline, the morphology of polyethylene (PE) single crystals can be considered well established. Nonetheless, many details of the structure are still disputed, particularly the nature of the fold surface [32] and the detailed structure of the crystalline core [33,34].

Much of the experimental evidence supporting the various proposed structures has involved measurements of dried-down single crystals. Recently, the thermal properties of PE single crystals maintained as suspensions in various liquids have been investigated [35–37]. Thermograms of dried lamellae are characteristically broad compared to those of low molecular weight materials due to reorganizational processes, distribution of crystallite sizes, and degrees of perfection within the crystal. However, the thermograms of single crystals suspended in liquid show a number of well-resolved, narrow peaks relative to the apparently single broad peak of the dried material. Explicit evidence for a higher degree of lateral crystalline order in crystals that have never been dried was obtained from x-ray line broadening measurements [33]. The calculated crystallite size of suspended crystals was found to correspond to a theoretical upper limit that is a consequence of chain obliquity. The measured crystallite sizes of dried-down material were determined to be significantly smaller.

These changes in physical properties upon drying suggests that many previous conclusions concerning the structure of single crystals need to be reexamined. For instance, much of the controversy concerning the nature of the fold surface, with apparently conflicting conclusions drawn from different types of experimental measurements, could be a consequence of a partial collapse of structure upon drying. The infrared spectrum of PE is sensitive to both crystalline order and the various conformations characteristic of amorphous regions and regular folds [20,38,39]. A comparison of the spectra of suspended and dried-down single crystals should therefore be revealing. However, a classic limitation of infrared spectroscopic studies of solutions or material suspended in a liquid has been the separation of the spectral contributions of the various components. With the recent introduction of computerized instruments, spectral subtraction of the liquid or solvent may now be readily accomplished [15]. Nevertheless, there are still some formidable experimental problems involved in obtaining the spectra of suspended PE single crystals [40].

The spectrum of polyethylene single crystals suspended in carbon tetrachloride is compared to that of the pure liquid in Fig. 22. Between 700–900 cm^{-1} effective total absorbance occurs due to the very strong CCl_4 fundamental modes. Even between 1700–900 cm^{-1} the presence of overtone and combination bands distorts the relative intensities of the polyethylene bending modes at 1473 cm^{-1} and 1462 cm^{-1}. However, in this region of the spectrum, the liquid bands can be removed by the spectral subtraction routine [15]. The difference spectrum obtained by subtracting the CCl_4 bands from the spectrum of suspended polyethylene single crystals is also presented in Fig. 22. The methylene bending modes near 1470 cm^{-1} appear very sharp and the relative intensities of the two bands, which have been used as an estimate of crystalline order [20], differed from the spectra of dried-down material. Unfortunately, a direct comparison of the spectra of suspended and dried crystals is complicated by either orientation effects or the mechanical disruption of structure, depending upon the method of sample

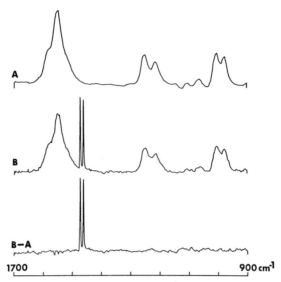

FIG. 22   (A) Infrared spectrum of CCl_4. (B) Infrared spectrum of polyethylene single crystals suspended in CCl_4. (B − A) Difference spectrum obtained by subtracting spectrum (A) from spectrum (B) so that the CCl_4 bands are reduced to the baseline. (From [31a].)

preparation. We will first briefly consider these effects before proceeding to a discussion of the differences between suspended and dried material.

The methylene bending and rocking modes of crystalline polyethylene are split into two components as a consequence of interchain interactions. The low-frequency component of each of these doublets has a dipole moment change parallel to the b axis of the unit cell, while for the higher frequency component this change is parallel to the a axis [41]. It has been demonstrated that a double orientation effect can influence the relative intensity of these bands in molded films [42]. In order to minimize such effects, previous studies of single crystals have been made on pressed mats of finely chopped material [43], or samples carefully dispersed into KBr [44]. However, the lamellae used in this study appeared labile to even the mildest mechanical attrition. In order to avoid effects due to mechanical disruption of structure, samples of dried single crystals were prepared simply by casting suspensions onto a KBr window. The effect of orientation on the relative intensities of the absorption bands was determined by comparing the spectra of films held perpendicular and at 45° to the incident beam [42].

The C–H stretching region of the spectrum of such a dried film is compared to that of crystals suspended in CCl_4 in Fig. 23. Since these bands are not split into a- and b-axis components by interchain interactions, no orientation effects were evident. However, the intensity of the CH_2 symmetric

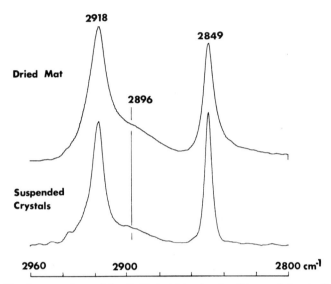

FIG. 23   A comparison of the C–H stretching bands of suspended polyethylene single crystals (CCl$_4$) with those of a dried mat obtained from the same suspension. (From [40].)

stretching at 2849 cm^{-1} decreases relative to the asymmetric stretching at 2918 cm^{-1} upon drying. Conversely, a weak shoulder near 2896 cm^{-1} is more prominent with respect to this latter band. At this time these spectral changes cannot be unambiguously interpreted. The 2897-cm^{-1} band has been assigned to either the b_{3u} component of the asymmetric CH stretching or a combination of the Raman-active and infrared-active methylene bending modes [45]. It is also possible that this band is associated primarily with an amorphous fraction. For such material a breakdown in selection rules would allow the Raman line near 2890 cm^{-1} to become active in the infrared.

The bands of the suspended material are significantly sharper than the dried crystals. This is the major difference observed for all the crystalline bands. The CH$_2$ rocking fundamentals of single crystals suspended in cyclohexane are compared to those of a dried mat and a melt-crystallized sample in Fig. 24. The successive increase in bandwidth is evident in this series. This is a slight orientation effect in the spectrum of the dried mat that results in a small change in the relative intensity of the 730 to the 720-cm^{-1} band with increasing angle of inclination of the sample to the beam. However, the band width remained constant in these spectra. The width at half-height for the C–H stretch and CH$_2$ bending and rocking modes of suspended crystals, dried mats, and melt-crystallized samples are presented in Table III. Drying results in an increase in this parameter of at least 10% for each of the bands.

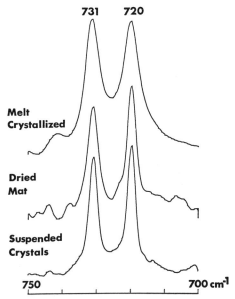

FIG. 24   A comparison of the CH_2 rocking bands of the spectra of: bottom, polyethylene single crystals suspended in cylohexane; middle, a dried mat of the same suspension; and top, a melt-crystallized film. (From [40].)

TABLE III

Comparison of Bandwidths of Crystalline PE

Infrared band (cm^{-1})	Band width at half height (cm^{-1})		
	Suspended crystals	Dried mat	Melt-crystallized film
2917	11.0	13.5	(Too thick)
2849	6.5	7.6	(Too thick)
1473	3.0	3.4	7.2
1462	3.0	3.4	7.2
731	3.0	3.8	6.0
720	3.4	4.0	7.0

There are a number of possible origins of this effect. Broadening of bands can occur through an increase in the proportion of amorphous material (kinks + jogs) in the crystalline core. However, the introduction of conformations other than the preferred all-trans sequence in a chain in the crystal results in an activation of the density of states for that chain such that the 720 and 1462-cm^{-1} bands are broadened asymmetrically on the high- and low-frequency sides respectively. The observed broadening of bands on drying suspended PE crystals is apparently symmetric, indicating that the concentration of such defects is small. The collapse of lamellae upon drying can involve tilting of chains, shear between planes, or probably a mixture of both [33]. Consequently, a distribution of interchain distances would be introduced (or broadened). This would in turn be reflected in the interchain interactions, resulting in a broadening of crystalline bands. However, even though the details of the structural changes leading to band broadening are to a certain degree open to interpretation, the relative sharpness of the crystalline bands of suspended material can only be due to a more perfect lateral register of chains in the crystal lattice.

The collapse and cracking of PE single crystals upon drying could also affect the structure of the fold surface, particularly if there is a regular fold conformation and adjacent reentry. Methylene wagging bands near 1350 cm^{-1} have been assigned to various gauche (G) and trans (T) sequences. Figure 25 compares the 1400–1300 cm^{-1} region of the spectra of single crystals suspended in CCl_4, a dried mat of the same material, and a melt-crystallized sample. The two prominent bands in the latter two spectra, at 1368 cm^{-1} and 1352 cm^{-1}, have been assigned to GTTG (or GTG) and GG conformations respectively [38,39]. However, a band at 1348 cm^{-1} is apparent in the spectrum of the suspended crystals, with a shoulder at 1352 cm^{-1}. Frequency shifts could occur through interactions between the

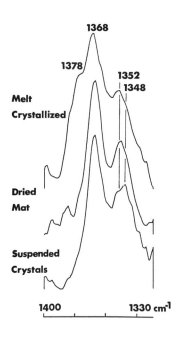

FIG. 25   The methylene wagging bands of polyethylene: bottom, suspended single crystals (CCl_4); middle, a dried mat of the same suspension; and top, a melt-crystallized film. (From [40].)

suspending medium and the polymer chain, particularly at the crystal surface. But, if this were the case, the 1368 cm^{-1} band should also shift.

A second possibility is that conformations present in an amorphous surface layer in suspended crystals differ from their dried counterpart. In support of this argument are the observations [46] that the long spacings of polyethylene single crystal mats change with the addition of various liquids, implying a reorganization of the fold surface. However, the methylene wagging bands of a dried mat swollen in CCl_4 in a liquid cell appeared very similar to that of the original dried mat. The swelling of the surface layers does not therefore greatly effect the distribution and types of conformation that give the characteristic infrared bands of dried material. Consequently, the 1348-cm^{-1} band can be correlated to a conformation present in suspended crystals that is at least partially destroyed in the drying and subsequent collapse of the lamellae. It is tempting to correlate this band also to a regular fold structure. In this respect a band near 1346 cm^{-1} has been detected in difference spectra of polyethylene single crystals [44]. The behavior of this band upon various treatments indicated an assignment to a conformation unique to a regular fold, possibly a distorted GG conformation. Although this correlation is tentative we suggest that it provides the most reasonable interpretation at this time.

In summary, from the above results we conclude:

(1)   The crystalline bands of dried polyethylene single crystals are significantly broader than the corresponding bands of material suspended in liquid. These results indicate a decrease in lateral crystalline order upon drying.

(2)   Drying also results in a change in the type and distribution of conformations associated with amorphous and fold structures, as determined by the observed frequencies of the methylene wagging bands.

ACKNOWLEDGMENTS

The authors would like to acknowledge the following scientists for major contributions to the work described in this paper.

Professor J. L. Koenig and W. W. Hart for the studies on carbon black filled elastomers. R. J. Petcavich for the investigation of the degradation of polyacrylonitrile. Professor I. R. Harrison and Maan–Shii S. Wu for the studies on polyvinylidene chloride. J. Zarian and D. F. Varnell for the PMMA/PVDF blend study. Professor I. R. Harrison and J. P. Runt for examination of suspended polyethylene single crystals.

In addition we would like to express our sincere thanks to Dr. Bruce Frushour and the Monsanto Company, the B. F. Goodrich Chemical Company, the Dow Chemical Company the Elastomer Chemicals Department of E. I. de Pont de Nemours & Company for their help in obtaining many polymeric materials.

The authors gratefully acknowledge the financial support of NSF Grant Nos. DMR 72-03292 A02, DMR 75-01254, and DMR 76-00887.

REFERENCES

1. W. W. Hart, P. C. Painter, J. L. Koenig, and M. M. Coleman, *J. Appl. Spectrosc.* **31**, 220–224 (1971).
2. F. Ciampelli and I. Manoviciu, *Gazz. Chim. Ital.* **91**, 1045–1051 (1961).
3. M. M. Coleman, P. C. Painter, D. L. Tabb, and J. L. Koenig, *J. Polym. Sci., Part B* **12**, 577–581 (1974).
4. P. C. Painter, M. M. Coleman, P. W. Wang, R. G. Jenkins, and P. L. Walker, Jr., *Fuel* (accepted for publication).
5. L. H. Peebles, Jr., "Degradation of Acrylonitrile Polymers." Technical Report #6, Office of Naval Research, Arlington, Virginia, 1976.
6. R. Olofson, private communication.
7. C. Y. Liang, F. Pearson, and R. Marchessault, *Spectrochim. Acta* **17**, 568–571 (1961).
8. D. L. Tabb and J. L. Koenig, *J. Polym. Sci., Part A-2* **13**, 1159–1166 (1975).
9. M. M. Coleman, *Polym. Prepr., Amer. Chem. Soc., Div. Polym. Chem.* **17**(2), 732–736 (1976).
10. R. R. Garrett, C. A. Hargreaves, II, and D. N. Robinson, *J. Macromol. Sci., Chem.* **4**, 1679–1703 (1970).
11. J. T. Maynard and W. E. Mochel, *J. Polym. Sci.* **13**, 235–250 (1954); *ibid.* **13**, 251–262 (1954).
12. M. M. Coleman, D. L. Tabb, and E. G. Brame, *Rubber Chem. Technol.* **50**(1), 49–62 (1977).
13. M. M. Coleman and E. G. Brame, submitted for publication.
14. C. A. Aufdermarsh and R. Pariser, *J. Polym. Sci., Part A* **2**, 4727–4733 (1964).
15. J. L. Koenig. *Appl. Spectrosc.* **29**, 293–308 (1975).
16. H. C. Bailey, *Rev. Gen. Caout. Plast.* **44**, 1495–1502 (1967).
17. I. R. Harrison and M. K. Louie, submitted for publication.
18. C. Cortili and G. Zerbi, *Spectrochim. Acta, Part A; Mol.* **23**, 285–299 (1967).
19. M. Kobayashi, K. Tashiro, and H. Tadokoro, *Macromolecules* **8**(2), 158–171 (1975).
20. S. Krimm and C. Y. Liang, *J. Polym. Sci.* **22**, 95–112 (1956).
21. S. Krimm, *Fortschr. Hochpolym.-Forsch.* **2**, 149–153 (1960).
22. S. Narita, S. Ichinohe, and S. Enomoto, *J. Polym. Sci.* **37**, 251–261 (1959); *ibid.* **37**, 263–271 (1959).
23. P. T. Hendra, J. R. Mackenzie, and P. Holliday, *Spectrochim. Acta Part A: Mol.* **25**, 1349–1354 (1969).
24. R. C. Reinhardt, *Ind. Eng. Chem.* **35**, 422–428 (1943).
25. C. S. Fuller, *Chem. Rev.* **26**, 143–167 (1940).
26. D. L. Tabb, J. L. Koenig, and M. M. Coleman, *J. Polym. Sci., Part A-2* **13**, 1145–1158 (1975).
27. S. Kraus, *J. Macromol. Sci., Rev. Macromol. Chem.* **7**(2), 251–319 (1972).
28. T. Nishi and T. T. Wang, *Macromolecules* **8**(6), 909–915 (1975).
29. J. S. Noland, N. N. C. Hsu, R. Saxon, and J. M. Schmitt, *Advan. Chem. Ser.* **99**, 15–28 (1971).
30. D. R. Paul and J. L. Altamirano, *Polym. Prepr., Amer. Chem. Soc., Div. Polym. Chem.* **15**, 409–414 (1974).
31. F. A. Bovey, F. C. Schilling, T. K. Kwei, and H. L. Frisch, *Polym. Prepr., Amer. Chem. Soc., Div. Polym. Chem.* **8**(1), 704–708 (1977).
31a. M. Coleman, J. Zarian, D. F. Varnell, and P. C. Painter, *J. Polym. Sci., Part B* **15,** 745 (1977).
32. B. Wunderlich, "Macromolecular Physics", Vol. I. Academic Press, New York, 1975.
33. I. R. Harrison and J. P. Runt, *J. Polym. Sci.* **14**, 317–322 (1976).

34. I. R. Harrison, A. Keller, D. M. Sadler, and E. L. Thomas, *Polymer* **17**, 736–739 (1976).
35. I. R. Harrison, *J. Polym. Sci.* **11**, 991–1003 (1973).
36. I. R. Harrison and G. L. Stutzman, *Anal. Calorimetry* **3**, 579–592 (1974).
37. I. R. Harrison, *J. Macromol. Sci. Chem.* **8**(1), 43–52 (1974).
38. R. G. Snyder, *J. Chem. Phys.* **47**, 1316–1360 (1967).
39. G. Zerbi, L. Piseri, and F. Cabassi, *Mol. Phys.* **22**, 241–256 (1971).
40. P. C. Painter, J. P. Runt, M. M. Coleman, and I. R. Harrison, *J. Polym. Sci. Polym. Phys. Ed.* **15**, 1647–1654 (1977).
41. S. Krimm, C. Y. Liang, and G. B. B. M. Sutherland, *J. Chem. Phys.* **25**, 549 (1956).
42. J. P. Luongo, *Polym. Lett.* **2**, 75–79 (1965).
43. J. L. Koenig and D. E. Witenhafer, *Macromol. Chem.* **99**, 193–201 (1966).
44. P. C. Painter, J. Havens, W. W. Hart, J. L. Koenig, *J. Polym. Sci. Polym. Phys. Ed.* **15**, 1223–1235 (1977).
45. J. R. Nielson and R. F. Holland, *J. Mol. Spec.* **6**, 394–418 (1961).
46. Y. Udagawa and A. Keller, *J. Polym. Sci., Part A-2* **2**(9), 437–451 (1971).

11

Reflection–Absorption Infrared Spectra of γ-Aminopropyltriethoxysilane Adsorbed on Bulk Iron*

F. J. BOERIO

DEPARTMENT OF MATERIALS SCIENCE
  AND METALLURGICAL ENGINEERING
UNIVERSITY OF CINCINNATI
CINCINNATI, OHIO

I. INTRODUCTION

The adhesion of organic polymers to inorganic substrates is of great importance in many areas, including adhesive bonding, coatings, and corrosion inhibition, and there has accordingly been great interest in developing methods for enhancing this adhesion. One of the most effective of these, especially when adhesion in the presence of liquid water or high humidity is required, is the use of organofunctional silanes as "coupling agents" to bridge the interface between an organic polymer and an inorganic substrate.

Numerous investigations of the structure and properties of organofunctional silanes adsorbed onto a variety of substrates have been reported and a number of theories have been offered to explain the mechanisms by which these compounds function as "coupling agents." The suggestion that silanes

* Portions of this chapter have been adopted from Boerio *et al.*, *J. Appl. Polym. Sci.* **22**, 203–213 (1978).

function by forming stable covalent bonds with both the resin and the substrate was first made nearly 30 years ago [1] in a report to the Naval Ordnance Laboratory. Little evidence for the formation of such bonds to the substrate has been reported but Koenig and Shih [2] obtained Raman spectra of vinyltriethoxysilane (VTES) adsorbed onto an E-glass fiber from aqueous solution and assigned a band near 1080 cm^{-1} to an Si–O–Si bond between the silane and the substrate. Koenig and Shih [2] also used Raman spectroscopy to follow the polymerization of methyl methacrylate in contact with E-glass fibers treated with VTES and found that some 30–40% of the vinyl groups copolymerized with the methyl methacrylate.

However, it is known, as indicated earlier, that organofunctional silanes are effective adhesion promoters when applied to metals. But, while metallosiloxane bonds may be formed between the adsorbed silane and the substrate, such bonds are not likely to be stable against hydrolysis [3] and it is in the presence of water or high humidity that silanes seem to have the most beneficial effect. Moreover, it has been shown that silanes may be strongly adsorbed onto gold, whose chemical inactivity makes covalent bonding with an adsorbed silane very unlikely [3]. In addition, it is known that silanes are sometimes effective adhesion promoters when the functional groups on the silane are, in principle, incapable of reaction with the resin [4].

Attempts to explain the "coupling" action of organofunctional silanes by a surface wettability theory have also been made. The first requirement for formation of an adhesive bond is that the liquid adhesive must wet the substrate. In order to obtain such wetting the surface tension of the liquid adhesive resin must be less than the critical surface tension of the substrate [5]. However, it is known that silanes as commonly applied to glass surfaces inhibit rather than enhance wetting of the surface by a resin [1]. Also, it has been shown that many of the most effective silanes have critical surface tensions less than 35 dynes/cm [3, 6–8] while the surface tensions of many liquid adhesives are considerably higher [1].

It has also been suggested that silanes form a deformable layer between the resin and the substrate, providing a mechanism for relief of stresses at the interface [9]. However, it has been shown that only a trace of silane, less than a monolayer, has a substantial beneficial effect on interfacial adhesion [10] and it has been argued [1] that such a small amount of silane could not possibly provide the suggested relaxation.

Most recently, Plueddemann [11] has suggested a mechanism by which hydrolyzed silanes are adsorbed onto the surfaces of hydrated minerals through the formation of hydrogen bonds. In the presence of water, these bonds can alternately be broken and then remade, thus permitting relaxation of stresses without loss of adhesion. This requires that the polymer at the interface be somewhat rigid, thus avoiding retraction of the resin from the substrate when hydrogen bonds are broken, maintaining the resin near the interface, and providing the opportunity for hydrogen bonds subsequently to be reformed.

None of these theories is completely satisfactory, however, and considerable uncertainty still exists concerning the mechanisms by which organofunctional silanes are adsorbed onto inorganic surfaces and subsequently enhance the adhesion of a polymer to the surface, especially in the presence of water or high humidity. This research was initiated to determine the mechanisms by which γ-aminopropyltriethoxysilane (γ-APS) functions as a coupling agent on iron.

The adsorption of γ-aminopropyltriethoxysilane (γ-APS) from solution onto bulk metals and glasses has been investigated by ellipsometry [12] and electron microscopy [13–15]. It has been determined that γ-APS is adsorbed on such substrates as a smooth continuous film but that agglomerated particles may be deposited on the films depending upon the method of sample preparation. Schrader [14] studied the adsorption of C^{14}-labeled γ-APS onto glass blocks and separated the adsorbed material into three fractions according to the ease of extraction from the blocks. The first fraction, about 97% of the adsorbed film, was extracted by water at room temperature. The second fraction was extracted by boiling water. But the third fraction, about one molecular layer, survived extraction in boiling water for 100 minutes. Lee [7] studied the wettability of films formed by γ-APS adsorbed onto microscope slides from methanol–water mixed solvents. The films were remarkably nonpolar and it was suggested that γ-APS was adsorbed with the electron pair on the nitrogen atom oriented toward or parallel to the glass surface.

Little is known about the structure of these films on the molecular level. Bascom [16] did, however, obtain ATR infrared spectra of γ-APS adsorbed onto germanium from cyclohexane and water. Spectra of films formed in these solvents were identical, indicating that the structures of the films were similar. Spectra obtained from films adsorbed from cyclohexane were, however, somewhat more intense, indicating that these films were thicker than those adsorbed from water. Bascom suggested that a broad band near 3000 cm^{-1} in the spectra of the adsorbed films was the result of hydrogen bonding of the amine groups which normally absorb near 3350 cm^{-1}. Little other information concerning the molecular structure of γ-APS adsorbed on bulk surfaces is available.

The purpose of this paper is to describe the use of reflection–absorption infrared spectroscopy to determine the structure of the films formed by γ-APS adsorbed onto bulk iron from aqueous solutions.

II. REFLECTION–ABSORPTION INFRARED SPECTROSCOPY

Infrared spectroscopy has been used successfully for some time to study adsorption on high surface area powders and it has long been known that this type of spectroscopy could, in principle, be used to investigate thin organic films formed on bulk metal substrates if the necessarily weak infrared absorption of such films could be suitably enhanced and detected. In the past, efforts have been made to do so by increasing the geometrical path length of the infrared rays in the film by making multiple reflections of the incident radiation at near-normal incidence from opposing samples as shown in Fig. 1. However, when light is incident on a highly reflecting metallic surface at near-normal incidence, the incident and reflected waves combine to form a standing wave having a node very near the surface of the metal [18]. Accordingly, the electric field has zero amplitude at the surface of the metal and it cannot interact with any molecules adsorbed on the surface. As a result, the sampling arrangement shown in Fig. 1 is not expected to be effective for small values of θ and, in fact, this has been shown experimentally [19].

Recently, however, Francis and Ellison [20] and Greenler [18] have reconsidered the problem of obtaining the infrared spectra of organic films formed on metallic surfaces. Their results have shown that only for radiation polarized parallel to the plane of incidence and making an angle of incidence that is only a few degrees less than 90° will the incident and reflected rays combine to establish a standing wave with appreciable amplitude at the surface of the metal that is capable of interacting with adsorbed molecules. This prediction of high absorption intensity for parallel polarized radiation and high angles of incidence is not based on the geometrical path length of the infrared radiation in the organic film. Accordingly this type of infrared spectroscopy is considerably different from conventional infrared spectroscopy, and Greenler and associates have suggested it be called reflection–absorption (R–A) spectroscopy [21].

Francis and Ellison [20] have shown that ΔR, the fractional change in reflectivity due to the presence of a film on a reflecting substrate, is given

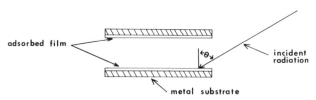

FIG. 1 Sample arrangement for obtaining multiple reflections at an angle θ from opposed coated mirrors. (From [17].)

approximately, for radiation polarized parallel to the plane of incidence, by

$$\Delta R = 1 - \frac{R}{R_0} = \frac{16\pi d \cos \theta}{\lambda}\left(\frac{A - B}{C}\right) \tag{1}$$

where

$$A = n_1 k_1 [k_4^2 \sin^2 \theta (k_4^2 \cos^2 \theta + 1) + n_1^4 (k_4^2 \cos^2 \theta + 1)] \tag{2}$$
$$B = n_4 k_4 (n_1^4 - n_1^2 \sin^2 \theta - n_1^6 \cos^2 \theta) \tag{3}$$
$$C = n_1^4 [k_4^2 \cos^2 \theta (k_4^2 \cos^2 \theta + 2) + 1] \tag{4}$$

and R and R_0 are reflectivity of the metal with and without the film, n_1 and k_1 are the refractive index and absorption constant of the film, n_4 and k_4 are the refractive index and absorption constant of the metal, θ is the angle of incidence, λ is the wavelength, and d is the thickness of the film.

Equation (1) is obviously a complex function of θ but it has been shown that ΔR has a maximum value at some particular value of θ that is obviously the optimum angle of incidence. The optimum angle of incidence varies somewhat with wavelength and with the optical constants n_1, k_1, n_4, and k_4 but generally lies near 88°.

For very thin films, ΔR will be small even if the spectra are obtained using the optimum angle of incidence. In such cases it is necessary to enhance ΔR in order to obtain usable spectra. This can be done by using scale expansion or by making multiple reflections so long as the total energy reaching the detector does not fall below $1/e$ or 37% of its initial value [22].

For angles of incidence less than about 80°, the second term on the right side of Eq. (1) is negligible and ΔR can be approximated [20] as

$$\Delta R = \frac{16\pi \, dk_1}{\lambda n_1^3} \frac{\sin^2 \theta}{\cos \theta} \tag{5}$$

Since

$$\frac{k_1}{\lambda} = \frac{\alpha}{4\pi} = \frac{2.303\varepsilon c}{4\pi} \tag{6}$$

where α is the absorption coefficient, ε is the molar extinction coefficient, and c is the concentration of absorbing species, ΔR may be written, for $0 \le \theta \le 80$, as

$$\Delta R = \frac{\varepsilon c d}{n_1^3}\left(\frac{9.212 \sin^2 \theta}{\cos \theta}\right) \tag{7}$$

III. EXPERIMENTAL TECHNIQUE

Samples for studying the adsorption of γ-APS onto iron by reflection–absorption infrared spectroscopy were prepared as follows. Coupons of Armco ingot iron (2 cm × 4 cm) were mechanically polished (final polishing was with 0.05-μm γ-alumina), repeatedly washed with distilled water, and then dried with a stream of nitrogen. Coupons prepared in this way were initially hydrophilic. However, after a short exposure to the laboratory atmosphere, the coupons became hydrophobic, apparently as a result of the adsorption of organic compounds. Analysis of polished coupons by Auger spectroscopy confirmed the presence of adsorbed organic compounds as well as adsorbed oxygen and nitrogen. No evidence for retention of the alumina polishing compound on the surface was found. Auger analysis also indicated the presence of an oxide layer approximately 30 Å thick.

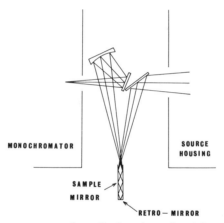

FIG. 2 Sampling arrangement for reflection–absorption infrared spectroscopy using multiple reflections from opposed sample mirrors. (From [17].)

In order to minimize contamination of the samples, coupons polished, washed, and dried as described above were immediately immersed in freshly prepared solutions of γ-APS. After removal from the solutions, the coupons were usually air dried and then mounted in a multiple reflection accessory as shown in Fig. 2. The accessory was then placed in the sample beam of a Perkin–Elmer 180 infrared spectrometer. Uncoated but otherwise identical coupons were mounted in a second reflection accessory which was then placed in the reference beam of the spectrometer, enabling all spectra to be recorded differentially. An AgBr wire grid polarizer, oriented to transmit only radiation polarized parallel to the plane of incidence, was placed in front of the entrance slit of the monochromator. The instrument was continuously purged with dried air to eliminate interfering absorption by atmospheric water vapor.

IV. RESULTS AND DISCUSSION

The spectrum shown in Fig. 3 was obtained using two reflections at 78° and 5× scale expansion from coupons exposed to an argon-purged, 1% aqueous solution of γ-APS at room temperature for 1 hour and then air dried. The intensity of the observed bands indicates that these films are several hundred angstroms in thickness and were probably formed by precipitation of polymerized γ-APS during the drying process rather than by adsorption. However, the presence of the strong band near 1575 cm^{-1} in the spectra of these thick films and its absence from spectra of the γ-APS monomer (Fig. 4) and from spectra of the polymer (Fig. 5) obtained by hydrolysis and condensation of γ-APS indicates that there are differences in structure between the film formed on iron coupons and the polymer.

The band near 1575 cm^{-1} undoubtedly corresponds to the NH$_2$ deformation vibration found near 1610 cm^{-1} in spectra of the monomer, shown in Fig. 4. There are several possible explanations for the observed shift in frequency from 1610 cm^{-1} to 1575 cm^{-1}. The band at 1575 cm^{-1} could be assigned to the asymmetric deformation of an NH$_3^+$ group formed by

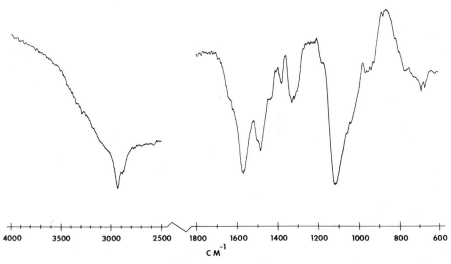

| 4000 | 3500 | 3000 | 2500 | 1800 | 1600 | 1400 | 1200 | 1000 | 800 | 600 |

C M^{-1}

FIG. 3   Reflection–absorption infrared spectrum of iron mirrors exposed to argon-purged, 1% aqueous solution of γ-APS at room temperature and then dried; two reflections at 78°, 5 × scale expansion. (From [17].)

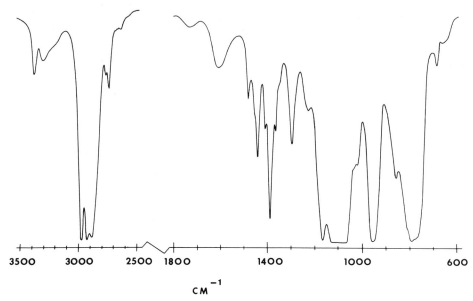

FIG. 4   Transmission infrared spectrum of γ-aminopropyl triethoxysilane. (From [17].)

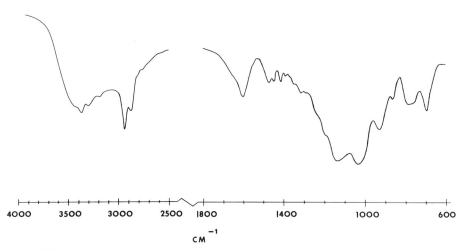

FIG. 5   Transmission infrared spectrum of γ-aminopropyltriethoxysilane polymerized by hydrolysis and condensation. (From [17].)

protonation of the amine. However, vibrations of this type are usually found at considerably higher frequencies. This mode has been assigned [23] to bands near 1613 cm^{-1} in propylamine hydrochloride and in butylamine hydrochloride and its assignment at 1575 cm^{-1} in γ-APS seems unlikely.

An alternative is to attribute the shift of the NH_2 deformation from 1610 cm^{-1} to 1575 cm^{-1} to hydrogen bonding. However, only relatively strong hydrogen bonding could explain a perturbation of the NH_2 deformation of the magnitude observed here. Strong hydrogen bonding of the O–H$\cdots$N type could also be expected for hydrolyzed γ-APS. The silanol group is rather strongly acidic and might be expected to form strong hydrogen bonds with the strongly basic amino group. However, little information is available in the literature describing the effects of such bonding on the deformation mode of the acceptor amine and it is not known if these bonds would be strong enough to produce the observed shift from 1610 cm^{-1} to 1575 cm^{-1}.

The spectrum shown in Fig. 3 is remarkably similar to spectra of ethylamine adsorbed on ethylammonium montmorillonite reported by Farmer and Mortland [24]. Ethylammonium montmorillonite has absorptions near 1617 cm^{-1} and 1510 cm^{-1} corresponding to the asymmetric and symmetric deformation modes of the NH_3^+ group. When ethylamine is adsorbed on ethylammonium montmorillonite, these absorptions are replaced by an absorption near 1590 cm^{-1}. This perturbation of the NH_3^+ deformation modes of the ethylammonium ion was attributed to strong N$\cdots$H$^+\cdots$N hydrogen bonding between ethylamine and ethylammonium. Hydrogen bonding of this type could also be obtained for γ-APS and might explain the occurrence of the NH_2 deformation mode near 1575 cm^{-1}.

The spectrum shown in Fig. 3 is also similar to spectra of coordinated primary amines. The NH_2 deformation mode in n-propylamine complexes with Cu(1)Cl is found near 1585 cm^{-1}, a downward shift in frequency of some 30 cm^{-1} from its position in the free amine [25]. Similar results have been reported for methylamine adsorbed on γ-alumina [26] and Farmer and Mortland [24] have shown that ethylamine coordinated to copper montmorillonite absorbs strongly near 1590 cm^{-1}.

It is felt that the most appropriate assignment of the 1575-cm^{-1} band is to the NH_2 deformation mode of amino groups having the nitrogen atoms coordinated to silicon atoms to form cyclic, inner complexes as suggested by Plueddemann [10]:

While complexes of silicon with nitrogen are not particularly common, several have been prepared and characterized, principally by x-ray diffraction [27,28]. An example is methyl (2,2′,3-nitrilodiethoxpropyl) silane having the structure [27]:

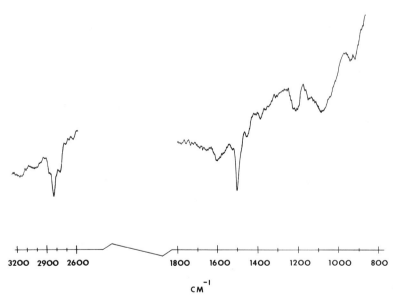

The coupons used to obtain the spectrum shown in Fig. 3 were subsequently soaked in toluene and then in ethyl acetate. It was expected that these solvents would dissolve monomer and low molecular weight polymer from the coupons. Spectra obtained from the coupons following this treatment were virtually identical to that shown in Fig. 3, indicating again that these thick films were composed mainly of polymerized γ-APS.

The same coupons were then washed in water at room temperature for 15 minutes and their spectra were again recorded as shown in Fig. 6. These spectra were obtained using two reflections at 78° and 20 × scale expansion.

FIG. 6   Reflection–absorption infrared spectrum obtained after iron coupons used to obtain Fig. 3 were washed for 15 minutes in water at room temperature; two reflections at 78°, 20 × scale expansion. (From [17].)

It may be observed that water has extracted most of the polysiloxane from the surface of the coupons, leaving only a thin film. Bands are observed near 2970 (shoulder), 2920, 2850, 1600, 1510, 1235, 1090, and 930 cm^{-1}. Observation of the band near 2970 cm^{-1}, assigned to Si–O–CH$_2$CH$_3$ groups, indicates that these thin films, which were probably formed by adsorption from solution, are composed of γ-APS that was only partially hydrolyzed. The band near 1090 cm^{-1} may be assigned to Si–O–Si linkages, indicating that the partially hydrolyzed γ-APS has condensed to form polysiloxanes.

The bands near 1510 cm^{-1} and 1600 cm^{-1} are assigned to the symmetric and asymmetric deformation modes of NH$_3^+$ groups. The corresponding modes for ethylammonium ions in ethylammonium montmorillonite are found [24] near 1617 cm^{-1} and 1510 cm^{-1}. Fripiat [23] et al. have assigned bands near 1613 cm^{-1} and 1510 cm^{-1} to NH$_3^+$ deformation modes in n-propylamine adsorbed on montmorillonite. These results indicate that initially γ-APS may have been adsorbed as low molecular weight (mainly dimers) internal zwitterions as suggested by Pluddemann and Erickson [1]:

The weakness of the asymmetrical NH$_3^+$ deformation mode near 1600 cm^{-1} relative to the symmetric deformation near 1510 cm^{-1} is thought to arise from an orientation effect. The spectra shown in Fig. 6 were, of course, obtained using radiation polarized parallel to the plane of incidence. As shown by Francis and Ellison [20] only the component of the dipole moment perpendicular to the surface will produce appreciable absorption in this case. For an isotropic film, this is of no particular consequence. However, for an anisotropic film, vibrations having transition moments parallel to the surface appear only weakly, while those having perpendicular transition moments appear with enhanced intensity. The intensity of the band near 1510 cm^{-1}, relative to that near 1600 cm^{-1}, indicates that the NH$_3^+$ group is oriented toward the surface, perhaps as the result of chelation with the iron surface. Fripiat [23] et al. have observed similar orientation affects in spectra of n-propylamine adsorbed on acid montmorillonite.

Subsequently, samples were prepared by immersing coupons in argon purged, 1% aqueous solutions of γ-APS at room temperature for 1 hr and immediately washing the coupons in distilled water following their removal from the solutions. The spectra shown in Fig. 7 were obtained using two reflections at 78° and 20× scale expansion from samples prepared in this way. It may be observed that these spectra are virtually identical to those in Fig. 6. These results confirm the earlier suggestions that the thick films characterized by the band near 1575 cm^{-1} are formed by precipitation of polymerized γ-APS during drying and that the thin films characterized by the band near 1510 cm^{-1} are formed by adsorption from solution. The observation of the band near 2970 cm^{-1} in Figs. 6 and 7 indicates that the adsorbed material is not completely hydrolyzed, while the absence of an absorption band near 2970 cm^{-1} in Fig. 3 indicates that the thick films deposited during drying are highly hydrolyzed.

In principle, the thickness of the films whose spectra are shown in Figs. 6 and 7 may be calculated from Eq. (7) provided ε, c, and n_1 are known. Unfortunately these quantities are not easily determined for an adsorbed film. Nevertheless, it is desirable to estimate the thickness of these films and that may be done as follows. Sandorfy and Jones [29] have suggested that the extinction coefficient for a chain of CH_2 groups is 75 1/mol-cm per CH_2 group. Thus for the propylamine group, $\varepsilon \simeq 225$ 1/mol-cm. The density of γ-APS is about 0.95 gm/cm^3 and for a molecular weight of 221 gm/mol, the density of propylamine groups is found to be about 4.3 mol/l. The refractive index for the adsorbed film can be approximated by the value for γ-APS monomer, $n_1 = 1.421$. Assuming $R_0 = 0.65$, ΔR for the band near 2920 cm^{-1} in Fig. 6 was found to be 0.0089, and d, the film thickness, was estimated at about 60 Å using Eq. (4). ΔR for the band near 2920 cm^{-1} in Fig. 7 was found to be 0.0025 and the film thickness was estimated at approximately 20 Å. Preliminary analysis of similarly prepared samples by ellipsometry have yielded comparable film thicknesses for these adsorbed films.

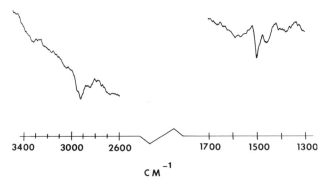

FIG. 7   Reflection–absorption infrared spectrum obtained from iron coupons exposed to argon-purged, 1% aqueous solution of γ-APS at room temperature, washed, dried; two reflections at 78°, 20× expansion.

V. CONCLUSIONS

Films having two distinct structures may be obtained on iron coupons immersed in dilute aqueous solutions of γ-aminopropyltriethoxysilane. Thin films having infrared spectra characterized by absorption bands near 1510 cm^{-1} and 2970 cm^{-1} are formed by adsorption from solution. The band near 1510 cm^{-1} is assigned to NH_3^+ groups, indicating that γ-APS is adsorbed as an internal cyclic zwitterion. The band near 2970 cm^{-1} is assigned to ethoxy groups indicating that these adsorbed films are incompletely hydrolyzed.

Thick films having infrared spectra characterized by an absorption band near 1575 cm^{-1} that is assigned to the NH_2 deformation mode for amino groups coordinated to silicon are formed by precipitation of polymerized γ-APS during drying of coupons removed from γ-APS solutions without rinsing. No absorption is observed near 2970 cm^{-1} in infrared spectra of these thick films, indicating that they are composed of highly hydrolyzed γ-APS. These thick films are readily dissolved by water to reveal the thin films described above.

Since the thick films are so weakly bound to the iron surface, it is unlikely that they play a significant role in enhancing the adhesion of polymers to the substrate. The thin, strongly bound films characterized by infrared absorption near 1510 cm^{-1} are undoubtedly dominant in that respect.

ACKNOWLEDGMENTS

The author wishes to express appreciation to the National Science Foundation (Grant GH-42679) and the American Iron and Steel Institute (Grant 66-345) for their financial support of this research. The many contributions of L. H. Schoenlein and J. E. Greivenkamp are also acknowledged.

REFERENCES

1. P. W. Erickson and E. P. Plueddemann, "Composite Materials" (L. J. Broutman and R. H. Krock, eds.), Vol. VI, pp. 1–29. Academic Press, New York, 1974.
2. J. L. Koenig and P. T. K. Shih, *J. Colloid Interface Sci.* **36**, 247–253 (1971).
3. W. D. Bascom, *J. Colloid Interface Sci.* **27**, 789–796 (1968a).
4. W. D. Bascom, "Composite Materials" (L. J. Broutman and R. H. Krock, eds.), Vol. VI, pp. 79–108. Academic Press, New York, 1974.
5. W. A. Zisman, *Ind. Eng. Chem.* **55**, 19 (1963).
6. W. D. Bascom, *Advan. Chem. Ser.* **87**, 38–62 (1968b).
7. L. H. Lee, *J. Colloid Interface Sci.* **27**, 751–760 (1968a).
8. L. H. Lee, *Advan. Chem. Ser.* **87**, 106–123 (1968b).
9. R. C. Hooper, *Proc. S.P.I. Conf. Rein. Plast. Div.*, *11th Sec. 8-B.* (1956).
10. E. P. Plueddemann, "Composite Materials" (L. J. Broutman and R. H. Krock, eds.), Vol. VI, pp. 201–216. Academic Press, New York, 1974.

11. E. P. Plueddemann, *J. Paint Technol.* **42**, 600–608 (1970).
12. D. J. Tutas, R. Stromberg, and E. Passaglia, *S.P.E. Trans.* **4**, 256 (1964).
13. S. Sterman, H. B. Bradley, *SPE (Soc. Plast. Eng.) Trans.* **1**, 224 (1961).
14. M. E. Schrader, "Composite Materials, Vol. VI" (L. J. Broutman and R. H. Krock, eds.), pp. 109–129. Academic Press, New York, 1974.
15. O. K. Johansson, F. O. Stark, G. E. Vogel, and R. M. Fleischmann, *J. Compos. Mater.* **1**, 278–292 (1967).
16. W. D. Bascom, *Macromolecules* **5**, 792–798 (1972).
17. F. J. Boerio, L. H. Schoenlein, and J. E. Greivenkamp, *J. Appl. Polym. Sci.* **22**, 203–213 (1978).
18. R. G. Greenler, *J. Chem. Phys.* **44**, 310–315 (1966).
19. H. L. Pickering and H. C. Eckstrom, *J. Phys. Chem.* **63**, 512–517 (1959).
20. S. A. Francis and A. H. Ellison, *J. Opt. Soc. Amer.* **49**, 131–138 (1959).
21. R. G. Greenler, *J. Catal.* **23**, 42–48 (1971).
22. R. G. Greenler, *J. Chem. Phys.* **50**, 1963–1968 (1969).
23. J. J. Fripiat, A. Servais, and A. Leonard, *Bull. Soc. Chim. Fr.*, pp. 635–644 (1962).
24. V. C. Farmer and M. M. Mortland, *J. Phys. Chem.* **69**, 683–686 (1965).
25. T. Ogura, T. Hamachi, and S. Kawaguchi, *Bull. Chem. Soc. Jap.* **41**, 892–896 (1968).
26. K. Hirota, K. Fueki, and T. Sakai, *Bull. Chem. Soc. Jap.* **35**, 1545–1548 (1962).
27. C. J. Frye, G. E. Vogel, and J. A. Hall, *J. Amer. Chem. Soc.* **83**, 996–997 (1961).
28. F. P. Boer and J. W. Turley, *J. Amer. Chem. Soc.* **91**, 4134–4139 (1969).
29. R. M. Jones and C. Sandorfy, "Techniques of Organic Chemistry" (W. West, ed.), Vol. IX, pp. 247–580. Wiley (Interscience), New York, 1956.

12

Dynamic Infrared of Polymers

R. P. WOOL

DEPARTMENT OF METALLURGY
    AND MINING ENGINEERING
UNIVERSITY OF ILLINOIS
URBANA, ILLINOIS 61801

W. O. STATTON *

MATERIALS SCIENCE AND ENGINEERING
THE UNIVERSITY OF UTAH
SALT LAKE CITY, UTAH

I. INTRODUCTION

Infrared (ir) spectroscopy is a well-established tool for polymer characterization and analysis. Recently, renewed interest in ir arose from deformation studies of polymers. It was shown by Zhurkov *et al.* [1], Gubanov [2],

* Present address: Kula, Hawaii.

Onogi and Asada [3], Uemura and Stein [4], Roylance and DeVries [5], and by Wool [6] that subtle but definite changes of ir band frequency and intensity of polymers in the stressed state revealed much information about molecular deformation processes and load bearing abilities of single chains. Typically, one observes an ir band deforming asymmetrically and shifting to lower frequencies. Other bands in the spectrum may increase or decrease in intensity without frequency shifting. These spectral changes, though often minute, can be used to obtain integrated responses of molecular, microscopic, and morphological interactions in terms of the samples deformation and thermal history.

The ir spectrum of a polymer contains bands which may be sensitive to stress induced changes in molecular orientation, conformational rotational isomeric states, molecular stress distributions, new end-groups due to bond rupture, structure, and morphology. We can use these bands to interpret the molecular mechanics of polymer deformation in terms of the macroscopic strain history. In this chapter, we will examine single chain and multichain polymer deformation phenomena by this dynamic infrared (DIR) method.

II. FREQUENCY SHIFTING

A. Band Deformation

The typical effect of an externally applied load on a stress-sensitive infrared band is shown schematically in Fig. 1. First, we observe a shift of the band maximum to lower frequencies. As discussed by Zhurkov et al. [1] and by Gubanov [2], the frequency shift, Δv, is proportional to the applied stress, σ, by

$$\Delta v = \alpha_x \sigma \tag{1}$$

where α_x is the experimentally determined proportionality constant.

Secondly, the "tail" or low-frequency wing of the band is related to the internal molecular stress distribution displacing minor vibrational bands of stressed chain segments nonuniformly to lower frequencies. Although the applied macroscopic stress may be uniform, the inhomogeneities at both molecular and microstructural levels result in a nonuniform stress distribution. Zhurkov et al. [7] suggest that the oscillator number distribution $F(v)$ along the frequency axis of the deformed band, $D(v)$, can be described by the convolution integral

$$D(v) = \int_{-\infty}^{\infty} F(\xi)U(v - \xi)\,d\xi \tag{2a}$$

or

$$D(v) = F(v) * U(v) \tag{2b}$$

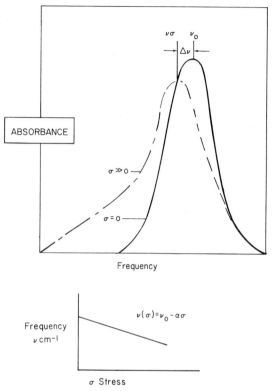

FIG. 1   Schematic of typically observed infrared band deformation and frequency shifts of "backbone" vibrations in stressed polymers.

where $U(v - \xi)$ is the shape of the undeformed band, ξ is a "dummy" variable, and the star implies convolution. $F(v)$ can be obtained by deconvolution knowing both $D(v)$ and $U(v)$ experimentally. Numerical deconvolution techniques for such processes have been discussed by Ergun [8], Kosobukin [9], Wool [6], and Mocherla [10]. The molecular stress distribution $F(\sigma)$ can be obtained by converting the frequency axis to a stress axis using Eq. (1), in which α_x has been interpolated to α_c, the frequency shift factor for a single chain.

In Eq. (2) it is assumed that $U(v)$, the elementary shape of the undeformed band, is independent of strain. This is not strictly true and other spectral aspects may contribute to the asymmetric band deformation as discussed by Wool [11]. However, it is considered that the convolution integral representation is a useful approximation for the deformed band.

DIR methods have the potential to unravel the intimate details of polymer deformation and lead to a more complete molecular description of the stressed solid state as a function of strain and thermal history. However, since this method is in its infancy, it seems prudent to take a closer look at the analytic tool before extensively using it. We will attempt to highlight the basic assumptions and approximations involved in application to molecular mechanics of polymers.

B. Mechanisms of Frequency Shifting

When a linear polymer chain is subjected to a tensile load along its axis, the most likely modes of chain deformation are bond stretching, internal rotation, valence angle changes, long-range disorder, and local defect disturbances. The principal individual contributions to frequency shifting from the above deformation modes are the following:

(a) Stress induced reduction of vibrational force constants due to anharmonic potential energy functions;
(b) Internal coordinate changes or simple elastic deformation of the chain skeleton; and
(c) Defect dependence of the density of states of vibrational modes.

Other mechanisms of frequency shifting can exist, in particular those associated with high-pressure induced frequency shifts discussed by Reynolds and Sternstein [12] for the high-pressure spectrum of poly(vinyl alcohol) and polyamides and by Wu and Shen [13] for the high-pressure spectra of amorphous polystyrene.

The reduction in force constants k depends on the potential energies of the strained segments associated with the vibration. For example, if we wished to describe the potential energy of a C–C bond by a Morse function, as in

$$U = D_0(1 - e^{-ar})^2 \tag{3}$$

then

$$k = \frac{\partial^2 U}{\partial r^2} = 2D_0 a^2 (2e^{-2ar} - e^{-ar}) \tag{4}$$

where D_0 and a are constants.

Thus k will vary as a function of the C–C bond length, r, so that k increases in compression and decreases in extension. The vibrational frequencies will change accordingly.

As demonstrated elsewhere by Wool [11], simple elastic changes of bonds and bond angles could also lead to frequency shifts without the necessity for force constant reduction. If one performs a normal mode analysis of strained polymers without changing the force constants, approximately half the bands shift to higher frequencies, some are unaffected by strain, and the remainder shift to lower frequencies.

Recently we examined the vibrational spectrum of isotactic polypropylene (PP) by DIR methods. The experimental results for each band were compared with the calculated normal mode analyses of Miyazawa [14] and Tadokoro et al. [15]. Many bands were found to shift to lower frequencies but none were found to shift to higher frequencies. Bands whose calculated vibrational modes consisted of C–C stretching components of the load-bearing segments of the chain were found to shift to lower frequencies. With the latter, the greater the contribution of C–C stretching to the total potential energy of the band, the greater was the observed shift. It was interesting to note that bands whose vibrational modes consisted solely of side-group vibrations did not show any tendency for frequency shifting. Several conformationally sensitive bands showed changes of intensity presumably due to variations in the concentration of oscillators associated with each conformation.

Thus, the dominant mode of frequency shifting appears to be the reduction of force constants caused by the weakening effect of tensile stress on the molecular framework. This effect is opposite to one's "intuitive" interpretation based on the analogy of strumming musical strings, where the acoustic frequency increases with stress.

The role of defects in the polymer chain may be difficult to quantize and relate to the vibrational spectrum. Polymer chains and lattices contain many defects such as crosslinks, chain ends, chain folds, etc., which can influence the vibrational spectrum. Zerbi et al. [16] used the negative eigenvalue theorem suggested by Dean and Martin [17] to investigate the effect of randomly induced conformational defects on the vibrational spectrum of linear polyethylene. His calculated density of state defect spectra were in reasonably good agreement with the high temperature spectrum of polyethylene. In the latter, it was found that in addition to the normal line broadening at elevated temperatures several bands shifted to higher and lower frequencies and deformed asymmetrically.

Thus, it could be argued that stress varies the concentration of natural defects and might cause some mechanically permissable but ir inactive bands to come into resonance. The new defect spectrum would distort the original and possibly influence the asymmetric band shape. If the defect mechanism is suspected, ir bands should be compared at several temperatures. Those which do not appreciably deform can, to a first approximation, be considered defect free.

The contribution of each mechanism to frequency shifting is unknown at the present time. Single-chain deformation involves changes in geometry and elastic constants, the exact form of which are unclear. Gubanov and Kosobukin [2] calculated the frequency shifts for strained isotactic PP, using the GF matrix method suggested by Wilson [18]. The quasielastic force constants were obtained from a calculated nonlinear potential strain energy for the single chain. Their results imply that many bands shift to high frequencies in addition to low frequencies and perhaps the extent of force constant reduction may have been underestimated. Their results are not in general agreement with the experimental results of Wool and Statton [11, 19].

At this point, we understand some general features of single-chain deformation but not the exact mode of geometry changes and the associated force constant variations.

C. The Frequency Shift Factor α

It is necessary to convert the frequency axis to a stress axis to facilitate computation of molecular stress distributions. This can be accomplished by substituting α_c in Eq. (1). Most previous works in this field have used α_x, the experimentally observed band shift, as a function of stress. This method is incorrect since α_x is both morphology and temperature dependent.

The morphology dependence of α has been convincingly demonstrated by Mocherla and Statton [20] for studies on poly(ethylene terephthalate) (PET). Figure 2 shows the dependence of the 976 cm^{-1} band of PET on

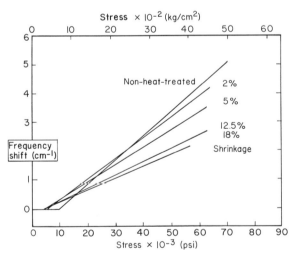

FIG. 2 The frequency of the 976 cm^{-1} band in PET is plotted as a function of stress for several samples with varying shrinkage from annealing treatments. When the shrinkage is 2%, 5%, 12.5%, 18%, and non-heat-treated, the corresponding valves of α (cm^{-1}/(kg/cm^2)) are 10.40 × 10^{-4}, 8.75 × 10^{-4}, 6.61 × 10^{-4}, 6.05 × 10^{-4}, and 11.66 × 10^{-4}.

annealing and strain history. It is found that α_x decreases with both annealing time and temperature. Slack annealing versus constant length annealing also reduces α_x. As discussed by Mocherla, α_x is a sensitive measure of the load bearing ability of the average polymer chain in the microstructure. As the modulus and strength of the sample decreases due to thermal or strain treatments, the shift factor decreases.

In Fig. 2, it is found that an initial stress level must be reached before shifting occurs. The latter is also morphology dependent and has been related

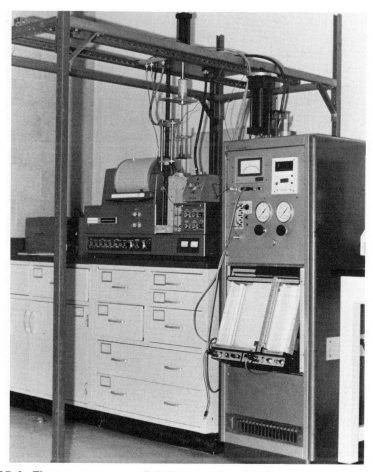

FIG. 3 The temperature-controlled dynamic infrared facility is shown. The piston-operated loading device with an oven around the sample area is inserted in the ir cavity of a Perkin–Elmer Model-621 spectrometer. The tensile loading and temperature control unit is on the right.

to load-bearing mechanisms via a morphological model of oriented semi-crystalline polymers by Mocherla and Statton.

For the 975 cm^{-1} band of PP, α_x may vary from .02 to .08 cm^{-1}/(kg/mm^2) depending on the degree of orientation along the principal stress axis [11].

Frequency shifting with stress is quite sensitive to the temperature of the sample [21]. The major effect of increased temperatures is to increase the amplitudes of vibration as in thermal expansion. This process further accommodates the reduction in force constants and aids in sample elongation with reduced fracture stresses.

Using the DIR experimental facility shown in Fig. 3, we examined a stress sensitive portion of the ir spectrum of highly oriented isotactic polypropylene as a function of temperature. Figure 4 shows the results of initial studies on the effect of temperature on the stress induced frequency shifts of the 941 cm^{-1} band. Both the 975-cm^{-1} and the 941-cm^{-1} bands showed significant variations of frequency with temperature. Since the 941-cm^{-1} band contains more backbone C–C stretching components than the 975-cm^{-1} band, it is expected to shift more. The 998-cm^{-1} "helix" band is relatively insensitive to stress, although some intensity variations occur.

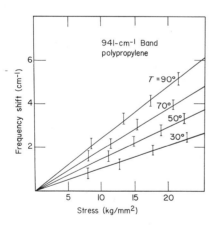

FIG. 4   The effect of temperature on stress induced frequency shifts of the 941-cm^{-1} band of polypropylene. The 941-cm^{-1} band contains C–C backbone stretching vibrations of the threefold PP helix.

An immediate consequence of frequency shifting worth noting is that the polymer chain modulus, E_c, is not a constant but depends on both the strain level and deformation temperature. From the above topics on frequency shifting with stress it can be deduced that E_c decreases in tension and high temperature and increases both in compression and with lower temperatures. These effects are discussed elsewhere [21a] with respect to E_c data evaluated by Sakurada et al. [22] using x-ray diffraction and by Rabolt and Fanconi [23] using Raman spectroscopic analyses of the longitudinal acoustic mode in linear helical polymers. Errors in measurements of E_c by x-ray can also occur by virtue of the nonhomogeneous stress distribution leading to similar phenomena such as the morphology and temperature dependence of α_x.

III. MOLECULAR ORIENTATION

A. Extensional Orientation

We use polarized ir methods to evaluate changes in molecular orientation. It is customary to define the dichroic ratio R as

$$R = A_{\|}/A_{\perp} \tag{5}$$

where $A_{\|}$ and $A_{\perp}$ are the absorbance intensities obtained with the incident electric vector of the beam polarized parallel and perpendicular, respectively, to the sample's extrusion or reference axis.

There are two mechanisms that can affect chain orientation measurements by DIR methods. One mechanism involves rotation or translation of entire chains or chain segments and thus affects the spatial distribution of chains with respect to some reference frame. For example, by applying a uniaxial stress, a randomly oriented polymer might achieve partial uniaxial orientation. Fraser [24], Beer [25], Stein [26], and others have shown how various spatial chain orientation functions can be obtained from polarized ir intensities and vice-versa.

Another mechanism called extensional ir orientation [28] considers changes in transition moment vectors **M** of a single chain under stress. In this case, the chain axis orientation remains unaltered during deformation but the angle θ between **M** and the chain axis depends on the strained chain geometry.

For singular vibrational modes in which **M** has one component vector directed along the skeletal bonds, R is obtained for helical polymers as

$$R = \frac{2(1 + \varepsilon)^2 \cos^2 \theta_0}{1 - (1 + \varepsilon)^2 \cos^2 \theta_0} \tag{6}$$

where ε and θ_0 are the chain strain and unperturbed transition moment angle, respectively.

Figure 5 shows the effect of strain on R as defined in Eq. (6) for several θ_0. With small θ_0 large increases in R can be obtained with increasing strain. For side-group vibrations of approximate perpendicular dichroism, the segmental orientation tends to decrease with strain.

When the components of **M** arise from coupled vibrations R is a more complex function of the ratios of the component vectors and the strain dependent helical parameters. Those results have been presented elsewhere by Wool [28].

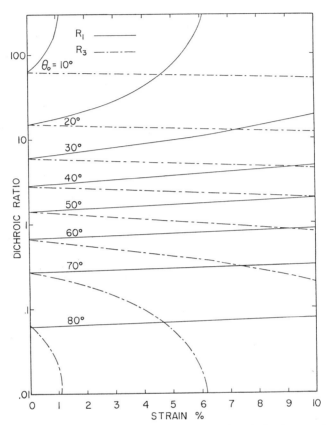

FIG. 5   The dichroic ratio is determined as a function of strain for the extensional orientation analysis of a simple helix. R_1 and R_3 were evaluated with respect to singular transition moments directed parallel and perpendicular to the skeletal bonds, respectively. (From [28].)

The extensional orientation mechanism provides a useful method of evaluating finite chain deformation and studies are continuing in this direction to evaluate changes in chain geometry with strain.

B. General Orientation and Stress

Several examples of the use of DIR in orientation studies have been presented by Zhurkov et al. [1], Vettegren and Novak [31], and by Wool and Statton [32]. These orientation studies fall into two classes depending on the stress sensitivity of the band. If the band does not show a frequency shift, then R is a measure of the integrated spatial orientation function coupled with the extensional orientation mechanism. When the latter effect is negligible, dichroic ratio changes as a function of deformation can then be related to actual changes in spatial distributions of chains. Equation (5) can be a good approximation to the true dichroic ratio if the macroscopic stresses are either reasonably low or the transition moment vector for the band is not too sensitive to the extensional orientation effect.

A more interesting and informative situation arises if the band shows a frequency shift distribution. Consider a rotationally symmetric distribution of chain axes, $f(\gamma)$, with respect to a reference cartesian coordinate set (x, y, z), where γ is the angle between each chain and the z-orientation axis. After Zbinden [33] we can calculate the polarized intensities I in the x, y, and z directions by relations of the type

$$I_x = \frac{1}{4\pi^2} \int_{\gamma=0}^{\pi/2} \int_{\phi=0}^{2\pi} \int_{\beta=0}^{2\pi} M_x^2 f(\gamma) \, d\beta \, d\phi \, d\gamma \tag{7}$$

where $M_x = \mathbf{TM}$ in which $\mathbf{T}$ is a transformation matrix converting components of $\mathbf{M}$ from chain coordinates to the cartesian coordinates and β and ϕ are projection angles of $\mathbf{M}$ in the chain and reference cartesian coordinates, respectively.

Under stress, the polarized bands will deform asymmetrically along the frequency axis as shown in Fig. 6 for both juxtaposed parallel and perpendicular polarized bands. The stressed band is now composed of overlapping bands A_p ($p = \|, \perp$) in which the chains associated with each A_p bear the same stress, σ_i, in the molecular stress distribution.

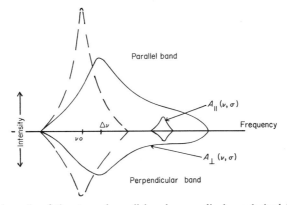

FIG. 6 Schematic of the stressed parallel and perpendicular polarized bands showing one of the minor components shifted to lower frequencies at some stress in the time-dependent molecular stress distribution.

We can now express the polarized stressed band as the convolution product

$$D_p(v) = A_p(v,\sigma) * U(v) \tag{8}$$

where $A_p(v,\sigma)$ is a function of the number, spatial distribution function, and chain stress, and can be expressed by relations similar to Eq. (7):

$$A_p(v,\sigma) = \frac{1}{4\pi^2} \iiint M_b^2 \sigma_i(\gamma)\, d\beta\, d\phi\, d\gamma \tag{9}$$

Here $\sigma_i(\gamma)$ is the spatial distribution of chains at the same stress σ_i and having the same transition moment angle θ_i.

The dynamic dichroic ratio of sets of chains at the same stress level in the molecular stress distribution can be obtained as

$$R(v,\sigma) = A_{||}(v,\sigma)/A_{\perp}(v,\sigma) \tag{10}$$

The $A_p(v,\sigma)$ can be obtained numerically at each instant by deconvolution of Eq. (9). Thus it is possible, at least theoretically to obtain a vast quantity of information regarding the coupling of the molecular stress distribution and the spatial orientation functions in terms of the macroscopic thermal and deformation history.

IV. APPLICATIONS

A. Stress Relaxation of Polypropylene

We wished to investigate the behavior of time-dependent molecular stress distributions and related phenomena during stress relaxation under constant strain in highly oriented viscoelastic PP. Initial results of such studies have been reported by Wool and Statton [32]. It was observed that an apparent accumulation of highly stressed chains developed during relaxation and eventually disappeared. Continuing these studies we were interested in the rapid intensity changes observed on the low-frequency (high-stress) wing of the 975-cm^{-1} band during the initial fast stress relaxation period.

The experiments were conducted on equipment similar to that shown in Fig. 3 and described previously by Wool and Statton [32]. Using an air-powered hydraulic piston we were able to approximate step function strain inputs on samples 2 inches long at their "beam" temperature of 30°C. Since the intensity changes can be rapid, it is necessary to obtain all parts of the band simultaneously to facilitate time-dependent molecular stress distribution analyses. This was accomplished by constant frequency testing on the

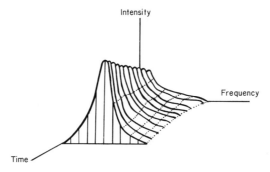

FIG. 7   Typical time-dependent intensity profiles of a stress-sensitive infrared band during stress relaxation. The dotted lines indicate the constant frequency intensities on the stress-sensitive wing of the band.

low-frequency wing of the 975-cm^{-1} band using a Perkin–Elmer Model-621 ir spectrometer. Figure 7 shows the typical constant frequency intensity vs. time curves (dotted lines) one can obtain on a stressed ir band contour during stress relaxation.

The experimental results of constant frequency tests are shown in Fig. 8 for the parallel polarized 975-cm^{-1} band during a step function strain input

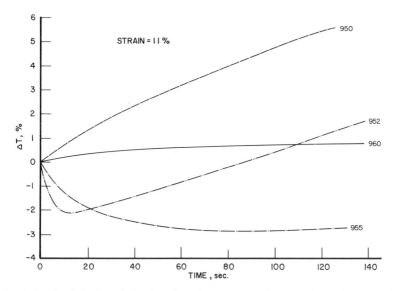

FIG. 8   The behavior of the time-dependent constant frequency intensities at various positions on the low-frequency side of the parallel polarized 975-cm^{-1} band during stress relaxation at step strain 11% in polypropylene.

of 11%. The change in transmission, ΔT, was determined with respect to the initial stressed intensities upon application of the step strain.

Most intensities appear to decrease as one would expect for a stress relaxation process. However, several increase for a period, maximize, and decrease later. This behavior suggests that when a number of highly stressed chains contribute to intensities at 950 cm^{-1}, the same intensity contribution should occur at a higher frequency, e.g., at 955 cm^{-1}, when stress relaxation has occurred. The intensity around 955 cm^{-1} can increase or decrease with time depending respectively on whether the number of stressed chains relaxing into that frequency region is greater than the number relaxing away from it. By this method, the number of highly stressed chains relaxing along the frequency axis can be monitored.

We examined molecular models to see whether the trends of the constant frequency data shown in Fig. 8 could be rationalized. Consider the model shown in Fig. 9. We have a chain length distribution, $\phi(l)$, between two rigid boundaries. Each chain is noninteracting with its neighbors, and has the ability to stress relax according to

$$\sigma(t) = G(t)\varepsilon_0 \tag{11}$$

where $G(t)$ and ε_0 are the stress relaxation modulus and step strain $\Delta l/l$ on each chain, respectively.

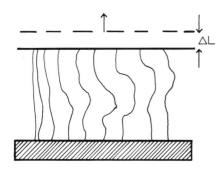

FIG. 9 A simple uniaxial molecular model showing a chain length distribution between rigid boundaries which can be deformed by an increment Δl. Each chain can stress relax independently of the others.

Assuming a Gaussian distribution for $\phi(l)$, we have

$$\phi(l) = 1/(s\sqrt{2\pi})\exp\{-[(l-\mu)^2/(2s^2)]\} \tag{12}$$

where s and μ are the standard deviation and mean chain length, respectively.

Upon application of a step Δl to the model as shown in Fig. 9, we can determine the number of chains at any stress σ and time t by

$$\phi(\sigma,t) = 1/(s\sqrt{2\pi})\exp\{-[(\mu\varepsilon_m/\sigma)G(t) - \mu]^2/(2s^2)\} \tag{13}$$

where the macroscopic strain input $\varepsilon_m = \Delta l/\mu$.

The intensity of the stressed ir band of the above model can be represented as the sum of the amorphous and crystalline components of the system:

$$D(v,t) = D_c(v,t) + D_a(v,t) \qquad (14)$$

The amorphous contribution arising from the chain length distribution can be expressed as the convolution product

$$D_a(v,t) = \phi(v,t) * U(v) \qquad (15)$$

in which the stress axis has been converted to a frequency axis using α_c.

If we let the stress in the rigid crystalline section be uniform and express the crystalline band fraction as a Lorentzian function, we have

$$D_c(v,t) = \frac{I_0}{1 + [(v_0(t) - v)/(\Gamma/2)]^2} \qquad (16)$$

in which I_0 and Γ are the band maximum intensity and halfwidth, respectively. The time-dependent frequency of the bands maximum intensity can be described by

$$v_0(t) = v_0 - \alpha_x \varepsilon_m G_m(t) \qquad (17)$$

where G_m is the macroscopic stress relaxation modulus.

Thus, the time-dependent shape of the deformed band for this model can be obtained by substituting for Eqs. (15–17) in Eq. (14). Equation (14) was evaluated by computer using data listed in Table I. The results are shown in Fig. 10 for the step strain input of 11%. Most of the calculated constant frequency intensities reach a maximum before decreasing, similar to the experimental results. The order of maximization of the intensities is shown in Table II for both model and polymer at the different strain levels. The latter are seen to be in reasonably good agreement despite the simplicity of the model and other possible sources of error discussed in the last section.

TABLE I

Parameters Used in Constant Frequency Intensity Calculations

Parameter	Value	Parameter	Value
α_x	0.021 cm^{-1} mm^2/kg	T	300°K
α_c	0.08 cm^{-1} mm^2/kg	E_c	42×10^4 kg/mm^2
$G'(t)$	$E_c e^{-\lambda t}$	s	60 Å
$G_m(t)$	Experimental value	μ	110 Å
ε_m	5, 8, 11%	k	1.987 cal/mol °K
λ	0.007	Γ_0	5 cm^{-1}

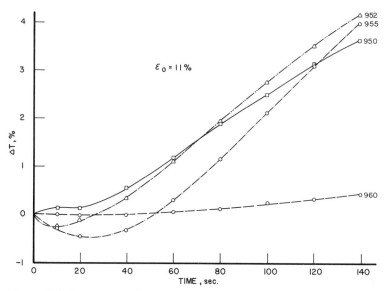

FIG. 10   Calculated constant frequency intensities for the molecular model during stress relaxation at 11% constant strain.

TABLE II

A Comparison of the Experimental and Theoretical Intensity Maximum for the Constant Frequency Tests of Polypropylene During Stress Relaxation at Several Step Function Strain Levels

5%	Observed:	$958 > 960 > 956 > 952 > 954$ cm^{-1}
	Calculated:	$958 > 956 > 960 > 954 > 952$ cm^{-1}
8%	Observed:	$956 > 954 > 958 > 952 > 960$ cm^{-1}
	Calculated:	$956 > 954 > 958 > 952 > 960$ cm^{-1}
11%	Observed:	$955 > 952 > 960 > 950$ cm^{-1}
	Calculated:	$955 > 952 > 960 > 950$ cm^{-1}

The above analysis provides for stress relaxation at all stress levels by a viscous nonfracture mechanism. Since stress relaxation can also occur by fracture mechanisms [34] we examined the spectrum of PP for new end-group frequencies. According to Veliev *et al.* [35] the primary radicals formed by bond rupture can react with absorbed oxygen or the parent chain to form new end groups. Possible end groups are the hydroxyl group at 3380 cm^{-1}, the ester group at 1742 cm^{-1}, the acid group at 1710 cm^{-1}, and a C=C group at 1645 cm^{-1}.

The latter frequencies were monitored during stress relaxation and it was concluded that a negligible amount of bond fracture occurred. The role of microcracks formed from rupture of van der Waals bonding was not determined in terms of the time-dependent molecular stress distributions. It is considered that such cracks play an important part in facilitating stress relaxation in addition to viscous processes [34].

The effects of previous strain histories on the mechanics of the present deformation were interesting and not always predictable. However, it was found that the chain length model was very useful in interpreting the results. For example, by slowly rising to a desired strain level, the number of highly stressed chains was always less than that of the faster rise time. By proceeding in small steps to 11% strain, the resulting molecular stress distributions during stress relaxation were very similar to that at approximately 8% strain if the latter had been applied as a single-step function.

The intensity variations were examined during a repeated loading–unloading (square wave) type strain deformation. It was found that the molecular stress distributions were essentially piecewise continuous while holding at each maximum strain interval. Thus, short unload times had little effect, whereas long unload times between cycles had an appreciable effect as the sample recovered. These studies indicate how samples tested at the same strain level can have widely varying molecular mechanics as a function of the strain history.

The effects of annealing on stress relaxation behavior were investigated. PP samples were annealed at 150°C under constant length conditions and then subjected to step function strain tests. The effects of annealing were to decrease the macroscopic stress relative to the unannealed sample. Consequently, very little activity of the low-frequency intensities were observed, indicating few highly stressed chains. Annealing decreased the number of load-bearing chains in the amorphous or tie chain regions and stress relaxation proceeded by extensive irreversible void formation as evidenced by "stress whitening" of the sample. The longer annealed samples exhibited a much narrower molecular stress distribution similar to a pure crystalline response, whereas the shorter annealed samples showed considerably more tie chain activity and a wider stress distribution.

B. Orientation Studies

The perpendicularly polarized 899-cm^{-1} and 809-cm^{-1} bands were examined during stress relaxation to determine the overall or average change in orientation. Their intensity behavior with time is shown in Fig. 11 at a constant strain of 11%. We observe for both bands that the intensity and thus the orientation increases rapidly during the first stages of the fast stress decay, then remains steady, and finally decreases with stress in the slow decay region. The net change in orientation is approximately zero.

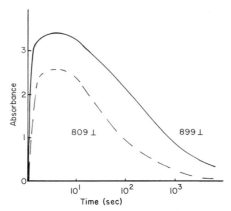

FIG. 11 Orientation changes during stress relaxation from 11% strain. The perpendicularly polarized intensities of the 809- and 899-cm^{-1} band (perpendicular dichroism) are plotted versus time during stress relaxation.

Similar effects can be observed for the 975-cm^{-1} band. We examined several frequencies on the low-frequency side of the 975-cm^{-1} band to obtain an approximate idea of the orientation behavior of the highly stressed regions. From the parallel and perpendicularly polarized constant frequency intensities, we determined the dynamic dichroic ratios during stress relaxation at 11% strain. The results are shown in Fig. 12 for the 950-, 955-, and 960-cm^{-1} constant frequencies. The stresses corresponding to the latter frequencies are 312, 250, and 187 kg/mm^2, respectively, based on $\alpha_c = 0.08$ cm^{-1} mm^2/kg.

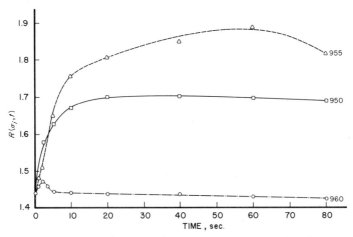

FIG. 12 Dichroic ratios of constant frequency intensities are plotted vs. time during stress relaxation of polypropylene. The frequencies were measured on the stress sensitive low-frequency side of the 975-cm^{-1} band.

The approximate (nondeconvoluted) orientation of all three stress levels rapidly increases in the first 10 seconds, i.e., in the fast decay region. The lowest stress maximizes first but does not achieve a very high level of orientation. (It should be noted that for the 975-cm^{-1} band the transition moment vector makes an angle of approximately 45° with the chain axis, resulting in a dichroic ratio of 2.0 for perfect axial orientation.) The higher stressed chains achieve a much higher orientation level before decreasing. During stress relaxation some chains relieve the stress by disorientation, while others increase their orientation. The disorientation process might arise from void formation [27], while the orientation process is the easiest method for a chain to accommodate itself to the stress direction. Also, highly stressed oriented amorphous chains could deorient as relaxation occurs through slippage or viscous processes. The latter is more likely in the slow stress decay region. Thus, while different stress levels show different orientation behavior as shown in Fig. 12, the average orientation behavior of the system will be determined by the majority of chains at lower stress levels as shown in Fig. 11.

These results have been obtained without separating the extensional orientation effect from the general chain axis orientation. However, due to the partial axial orientation of the sample and the similar orientation behavior of different bands, it is considered that the results are reasonably indicative of the traditional orientation measurements.

A similar series of DIR experiments were conducted to determine the molecular mechanics of constant load or creep of viscoelastic PP. Since creep is the opposite macroscopic deformation to stress relaxation in a viscoelastic sense, it might be expected that the molecular mechanics of both processes would also be converse. This was found to be correct, as shown by the general comparison of the molecular mechanics of stress relaxation and creep in Table III.

TABLE III

*Major Molecular Differences Resulting from Converse Macroscopic
Deformation Modes of Viscoelastic Polypropylene*

Stress relaxation	Creep
1. The number of highly stressed bonds decreases with time.	The number of highly stressed bonds increases with time.
2. The number of intermediately stressed bonds increases with time.	The number of intermediately stressed bonds decreases with time.
3. Helix bands decrease.	Helix bands increase.
4. Orientation decreases.	Orientation increases.
5. Small number of fractured bonds.	Large number of fractured bonds.

V. CONCLUSIONS

It is considered that dynamic infrared experiments can be extremely useful in elucidating the molecular mechanics of polymer deformation processes. A single experimental technique has the potential of obtaining single-chain, multichain, and morphological deformation behavior as a function of the macroscopic strain and thermal history. This method is in its infancy and requires much work and future study to realize its full potential.

ACKNOWLEDGMENTS

The authors are grateful to the National Science Foundation, GRANT ENG-76-09627, and the Research Foundation of the City University of New York for their financial support of this work. We are indebted to Dr. K. Mocherla for his consultation and permission to reproduce parts of his unpublished Ph.D. thesis.

REFERENCES

1. S. N. Zhurkov, V. I. Vettergren, V. E. Korsukov, and I. I. Novak, *Fracture 1969: Proc. 2nd Int. Cont. Fracture*, p. 545. Chapman & Hall, London, 1969; S. N. Zhurkov and V. E. Korsukov, *J. Polym. Sci.* **12**, 385 (1974).
2. A. I. Gubanov, *Mekh. Polim.* **3**, 771 (1967); A. I. Gubanov and V. A. Kosobukin, *Mekh. Polim.* **4**, 586 (1968).
3. S. Onogi and T. Asada, *J. Polym. Sci., Part C* 1445, (1967).
4. Y. Uemura and R. S. Stein, *J. Polym. Sci., Part A-2* **10**, 1691 (1972).
5. D. K. Roylance and K. L. DeVries, *J. Polym. Sci., Part B* **9**, 443, (1971).
6. R. P. Wool, Ph.D. thesis, Univ. of Utah, Salt Lake City, Utah, 1974.
7. S. N. Zhurkov, V. I. Vettegren, V. E. Korsukov, and I. I. Novak, *Fiz. Tverd. Tela* **2**, 290 (1969).
8. S. Ergun, *J. Appl. Crystallogr.* **1**, 19 (1968).
9. V. A. Kosobukin, *Sov. Phys.—Solid State* **14**, 2246 (1973).
10. K. Mocherla, Ph.D. thesis, Univ. of Utah, Salt Lake City, Utah, 1976.
11. R. P. Wool, *J. Polym. Sci.* **13**, 1795. (1975).
12. J. Reynolds and S. S. Sternstein, *J. Chem. Phys.* **41**, 47 (1964).
13. D. K. Wu and M. Shen, *J. Macromol. Sci., Phys.* **7(3)**, 549 (1973).
14. T. Miyazawa, "Stereochemistry of Macromolecules" (A. D. Ketley, ed.). Dekker, New York, 1968.
15. H. Tadokoro, T. Kitazawa, S. Nozadura, and S. Murahashi, *Bull. Chem. Soc. Jap.* **34**, 1209 (1961).
16. G. Zerbi, L. Pisen, and F. Cabassi, *Mol. Phys.* **22**, 241 (1971).
17. P. Dean and J. L. Martin, *Proc. Roy. Soc., Ser. A* **259**, 409 (1960).
18. E. B. Wilson, *J. Chem. Phys.* **7**, 1047 (1939).
19. R. P. Wool, and W. O. Statton, *Polym. Prepr. Amer. Chem. Soc., Div. Chem. Polym.*, **17**, 749 (1976).
20. K. Mocherla and W. O. Statton, *Symp. High Polym. Phys., Seoul, Korea, 1975, 1.*
21. R. P. Wool, paper presented at *MACRO I.U.P.A.C. Meet., Dublin, 1977.*
21a. R. P. Wool and W. O. Statton, "Recent Advances in Fiber Science" (F. Happey, ed.), Vol. 2, Chapter 32. Academic Press, New York, 1978.

22. I. Sakurada, T. Ito, and K. Nakamae, *J. Polym. Sci.* **15**, 75 (1966); I. Sakurada and K. Kahi, *J. Polym. Sci. Part C* **31**, 57 (1970).
23. J. F. Rabolt and B. Fanconi, *J. Polym. Sci., Part B* **15**, 12 (1977).
24. R. D. B. Fraser, *J. Chem. Phys.* **21**, 1113 (1958). W. Glenz and A. Perterlin, *J. Polym. Sci.* **9**, 1191 (1971).
25. M. Beer, *Proc. Roy. Soc. Ser. A* **236**, 136 (1956).
26. R. S. Stein, *J. Polym. Sci.* **31**, 327 (1958).
27. R. P. Wool, *J. Polym. Sci.* **14**, 603 (1976).
28. R. P. Wool, *J. Polym. Sci.* **14**, 1921 (1976).
29. R. P. Wool, paper No. 21S presented at *69th Annu. Meet. A.I.Ch.E., Chicago*, 1976.
30. R. P. Wool, *Int. J. Fracture*, in press.
31. V. I. Vettegren and I. I. Novak, *J. Polym. Sci.* **11**, 2135 (1973).
32. R. P. Wool and W. O. Statton, *J. Polym. Sci.* **12**, 1575 (1974).
33. R. Zbinden, "Infrared Spectroscopy of High Polymers." Academic Press, New York, 1964.
34. R. P. Wool, *Polym. Prepr. Amer. Chem. Soc., Div. Polym. Chem.* **18**, 111. (1977).
35. S. I. Veliev, V. I. Vettegren, and I. I. Novak, *Mekh. Polim.* **3**, 443 (1970); S. I. Veliev, V. E. Korsukov, and V. I. Vettegren, *Mekh. Polim.* **3**, 387 (1971).
36. R. P. Wool, unpublished data.

13

Characterization of Polymer Deformation and Fracture

DAVID K. ROYLANCE

DEPARTMENT OF MATERIALS SCIENCE AND ENGINEERING
MASSACHUSETTS INSTITUTE OF TECHNOLOGY
CAMBRIDGE, MASSACHUSETTS

I. KINETIC MODELS OF POLYMER FRACTURE

Although several useful phenomenological models of polymer fracture have been developed over the years, the materials scientist is compelled to seek out analyses that treat the fracture process from first principles. As an illustration of the sort of rate–process treatment often attempted, consider a thermally activated, stress-aided first-order process in which the concentration, N, of unbroken main-chain bonds decreases with time according to

$$-dN/dt = KN \tag{1}$$

207

where the rate constant K is given by the absolute theory of reaction rates [1] as

$$K = (kT/h)\exp[-(\Delta G^\dagger - \gamma\psi)kT] \tag{2}$$

Here k is Boltzmann's constant, T is the absolute temperature, h is Planck's constant, $\Delta G^\dagger$ is the activation free energy for bond dissociation, γ is the activation volume, and ψ is the local stress on the bond. Introducing the activation enthalpy and entropy through the familiar relation $\Delta G = \Delta H - T\,\Delta S$, Eq. (2) can be written

$$K = K_0\exp[-(\Delta H^\dagger - \gamma\psi)/kT] \tag{3}$$

where

$$K_0 = (kT/h)\exp(\Delta S^\dagger/k) \tag{4}$$

In Eq. (3) the temperature dependence of the preexponential term is usually neglected in comparison with the much stronger temperature dependence in the exponential term.

To proceed, one must now assume some functional relationship between the molecular stress ψ and the macroscopically imposed stress σ, and then integrate Eq. (1) to the limit of zero surviving bonds in order to predict the rupture time for the polymer. A particularly simple approach would be to assume that the molecular stress is uniform and unchanging, and equal to the imposed stress. The rate constant K is then truly constant, so that Eq. (1) integrates to

$$N = N_0\exp(-Kt) \tag{5}$$

The mean time to rupture of a given bond in the above kinetic process is easily shown [2] to be just the reciprocal of the rate constant K. If one now claims that the time to rupture of the stressed solid is equal to the average time for bond scission, one obtains the useful and well known Zhurkov equation [3]

$$\tau = \tau_0\exp[(U - \gamma\sigma)/kT] \tag{6}$$

Here $\tau_0 = 1/K_0$ is close to the period for atomic bond-pair vibrations and the apparent activation energy, U, for the fracture process is related to the activation enthalpy for bond dissociation $\Delta H^\dagger$.

The above derivation is wholly unrealistic, of course; one is certain that the molecular stress must vary widely within the material depending on the local morphology, and that it will change with time due to molecular rearrangements and scissions under stress. The parameters τ_0, U, and γ can be chosen so as to bring Eq. (6) into agreement with creep-rupture data for a wide variety of polymers, metals, and ceramics, but the parameters so chosen then reflect not only the kinetic properties of the interatomic bond but the morphology of the material as well. The activation volume is particularly structure sensitive. It is a measure of the effectiveness of the stress in overcoming the activation barrier for scission, and is thus strongly related to the stress-concentrating properties of the internal defect structure.

Even though Eq. (6), as well as many other considerably more elaborate rate–process fracture models, can be brought into line with experimental observation, one often feels that this is more a measure of the obliging nature of the exponential rate equations with their several adjustable parameters than any inherent realism of the model. All such models must involve certain assumptions concerning the nature of the bond dissociation process and distribution of molecular stress which until recently have been impossible to verify, and this has prevented the wide acceptance of any of these treatments. Only in the last decade has the missing element—direct experimental observation of the molecular fracture processes—been available to guide the development of more realistic models. Electron spin resonance (ESR) spectroscopy has proven particularly valuable in this regard.

II. ESR OBSERVATIONS OF STRESS-INDUCED BOND RUPTURE IN POLYMERS

A. Fundamentals of ESR Spectroscopy

As the theory of electron spin resonance spectroscopy is extensive and involved, only a brief description of its essential principles can be given here. The book by Poole [4] includes an extensive bibliography of the pertinent literature. ESR is a form of microwave absorption spectroscopy in which transitions are induced between the Zeeman energy levels arising from the interaction of an assemblage of paramagnetic electrons with an externally applied magnetic field. Upon application of a magnetic field of strength H, the normally degenerate spin of an electron is split in two levels separated by an energy difference ΔE given by

$$\Delta E = g\beta H \tag{7}$$

where g, the spectroscopic splitting factor, is ~ 2.0023 for a free electron, and the Bohr magneton β is a fundamental measure of the electron's magnetic character. At thermal equilibrium the electron populations of these two energy levels are given by Boltzmann statistics as

$$N^+/N^- = \exp(-\Delta E/kT) = \exp(-g\beta H/kT) \tag{8}$$

At ordinary temperatures $g\beta H \ll kT$, producing a slight excess of electrons in the lower energy state.

The Planck–Einstein relation states:

$$\Delta E = h\nu = g\beta H \qquad (9)$$

Transitions between the two energy levels may be induced by incident radiation of frequency $\nu = g\beta H/h$, which is right circularly polarized along the magnetic field axis. These transitions occur in either direction with equal probability but, since the lower state is more densely populated, the upward transitions outnumber the downward. If thermal equilibrium as given by Eq. (8) is maintained by spin–lattice relaxations, some electrons in the upper state may relax to the lower state by giving their excess energy to their surroundings. This produces a net energy loss from the incident radiation, which may be detected and displayed by suitable instrumentation. In practice, ESR spectrometers display energy absorption as a function of magnetic field strength H with the frequency ν held constant.

In order to exhibit paramagnetic resonance absorption, a solid specimen must contain electrons that are unpaired, i.e., that are not interacting with neighboring electrons to such an extent as to be subject to the restrictions of the Pauli exclusion principle. When homolytic scission of a covalent bond in a polymer takes place, the two electrons that had been paired in the bond become uncoupled, forming two free radicals that can be detected and identified by ESR spectroscopy. The study of free radical chemistry is a very active research area and ESR has become a standard analytic technique in this field. Many different types of degradation processes have been investigated extensively: thermal degradation; ultraviolet irradiation; neutron bombardment; γ irradiation; etc.

The single resonance peak predicted by Eq. (9) is often split into a series of peaks known as hyperfine structure, due to the electron's ability to interact with the magnetic moments of nearby nuclei as well as with the externally applied field. Since this hyperfine structure is characteristic of the particular active nuclei, it provides a "fingerprint" of a free radical's local chemical environment and can therefore be used to identify the radical.

B. Studies on Drawn Fibers

1. General

A principal requirement for the successful detection of stress-generated radicals is that they be created in sufficient numbers during the fracture process to exceed the sensitivity limit of the ESR spectrometer. In general, this requirement is not met. Polymers with amorphous or spherulitic structures tend to fracture in a localized manner, usually near an internal or surface imperfection that serves as a stress riser. The number of molecular chains passing through a given cross section is approximately 10^{14} cm^{-2}, and the number of scissions created by the passage of a planar crack through the section of a typical specimen is generally so low as to escape detection by ESR. Large radical concentrations can be achieved in most polymers by grinding them so as to produce a large amount of fracture surface, but the stress states operative during grinding are very complex and not conducive to analytic modeling.

To date, the material most amenable to ESR analysis during uniaxial tension has been found to be drawn polyamide fibers. The morphology of these materials is one of more or less aligned fibrils, these fibrils being columnar series of relatively rigid folded-chain lamellar fragments connected by molecular tie chains of various lengths [5]. The crystallites act as crack stoppers in a manner analogous to the reinforcing fibers in composite materials, so that the fracture damage occurs globally throughout the specimen rather than being restricted to a single locality. In addition, the extensive hydrogen bonding present in polyamides serves to inhibit interchain separation, forcing the crack to pass through the relatively weak tie-chain regions and cause covalent bond scission.

Figure 1 is a typical presentation of the ESR spectra generated by a series of stepwise-increasing stresses applied to drawn nylon 6 fibers. ESR spectrometers commonly employ phase-sensitive detection as a means of improving their sensitivity, and these spectra represent a derivative presentation of the microwave absorption. The number of radicals contributing to a given spectrum can be computed by a double integration of the derivative presentation, and comparison of the result with that of a standard such as diphenylpicrylhydrazine containing a known and stable number of radicals. The

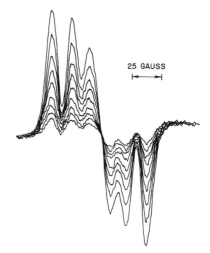

25 GAUSS

FIG. 1   Growth of ESR spectra during stepwise loading of drawn nylon 6 fibers Specimen was predesiccated and tested at 40°C in a dry nitrogen atmosphere. (From [13].)

number of radicals created during fracture depends on time, temperature, stress, and many other factors, but the approximate number of spins observed at the time of final rupture of nylon fibers at room temperature is approximately 10^{18} spins/cm^3.

The spectrum of Fig. 1 is identical to that observed in nylon after other forms of degradation, whether thermal, radiative, or mechanical. It has been shown [6] to be quartet arising from the radical —CO—NH—ĊH—CH$_2$— (~ 26 gauss hyperfine splitting with the one α and the two β protons), and a superimposed singlet due to —CH$_2$—ĊOH—NH—CH$_2$—. These radicals are clearly not at the original scission site, but at more stable sites along the chain to which they have migrated. Originally formed radicals in nylon have been observed after grinding at cryogenic temperatures, but the tendency of radicals to undergo transfer and other secondary reactions after fracture is perhaps the most serious limitation of ESR's use in polymer fracture studies.

Unpaired electrons are highly unstable, and seek to form bond pairs either by combination with other radicals or by reaction with other chemical groups. Radical decay occurs by a relatively rapid first-order kinetic process in the presence of excess oxygen, but can be shifted to a much slower second-order process by testing in an inert atmosphere. A typical data acquisition scheme involves determining the appropriate rate constants, and then correcting the observed radical count for any decay that may have occurred. The decay kinetics should be measured with the specimen under load, since the migration and combination mechanisms are influenced by specimen stress.

Data acquisition is made much more convenient by avoiding the need to scan the full ESR spectrum at various times during loading. One often finds that the height of a single selected peak on the derivative presentation is proportional to the second integral of the spectrum, so that the spectrometer is simply tuned to sit on the chosen peak during loading. This peak is then converted to spin count by the experimentally determined proportionality factor. Computerized data handling techniques make the entire process much more efficient.

2. The Lloyd–DeVries Bond Rupture Model

Most of the earlier rate–process polymer fracture models predict an accelerating rate of bond scission under constant stress, since scission of a particular bond serves to increase the molecular stress borne by its neighbors. The ESR observations are in direct contradiction to this prediction, however. As seen in Fig. (2), the rate of bond rupture in nylon fibers under constant stress decreases monotonically with time. DeVries, Lloyd and Williams [7] interpreted the decreasing rate of bond rupture in creep tests in terms of

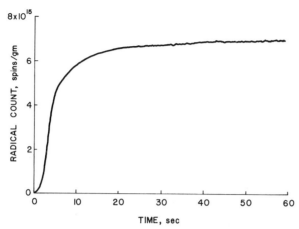

FIG. 2   Free radical production during a constant stress (690 MPa) test of drawn nylon fibers at room temperature. (From [13].)

a tie-chain contour length model, in which those molecular tie chains that are too short to accommodate a given specimen strain rupture, transferring their load to less critically stressed chains.

Pursuing this idea, Lloyd [8] used the ESR spectrometer to count the percentage of total fracture radicals generated during successively larger increments of specimen strain, thus developing a criticality histogram for the material (Fig. 3). The width of this histogram is an important structural parameter, having a strong influence on the ultimate tensile strength of the material. It is a function of the specimen's thermomechanical processing history, and is also influenced by the specimen testing temperature. Based

FIG. 3   Radical production histogram for nylon 6 fibers from step-strain tests at room temperature. (From [8].)

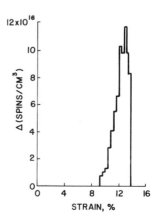

on ESR measurements of these histograms, Lloyd developed a rate–process model in which the distribution is considered as a large number of discrete subsegments and a rate equation similar to Eq. (1) written for each of them. A computer is then used to tally the extent of bond scission in each subsegment at each of a number of discrete time increments. The resulting model contains a certain number of adjustable parameters chosen so as to bring the predictions of bond rupture into line with selected ESR observations, but once chosen were able to predict the bond rupture kinetics for a wide variety of arbitrary load–temperature histories.

Lloyd's model is noteworthy principally in that it combines the Eyring description of covalent bond dissociation kinetics with an experimentally determined measure of the internal stress distribution of the material. Given the complex nature of even the relatively simple fiber morphology, one is not surprised that a computer treatment of the kinetic equations is necessary. Although developed principally as a means of predicting bond rupture during the loading process, the Lloyd model is also capable of predicting the final specimen fracture. One simply follows the scissions numerically until no unbroken bonds remain in any of the distribution subsegments. The present author recently employed Lloyd's model, along with the numerical parameters chosen by him to fit his "nylon 6 #1" material, to predict creep-rupture times for a series of constant stresses and temperatures. These data are plotted in Fig. 4, where it is seen that they show the linear variation of log lifetime with applied stress predicted by Eq. (6). Those data permit a selec-

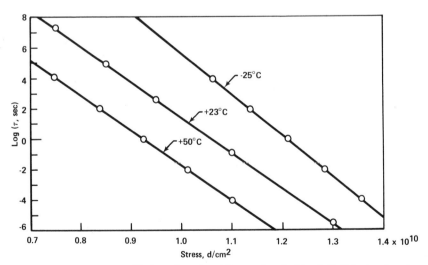

FIG. 4   Predicted creep lifetimes for nylon 6 fibers using the Lloyd–DeVries bond rupture model.

tion of the parameters τ_0, U, and γ in Zhurkov's equation, and the resulting expression is then capable of predicting final rupture times for a variety of loading histories.

C. Other ESR Fracture Studies

Although fiber studies of the sort described above have been most influential in leading to improved understanding of molecular fracture processes, much useful though more qualitative information has been obtained by ESR experiments on less tractable systems. Recently reviewed by DeVries and Roylance [9], these have included studies of grinding and abrasion of spherulitic polymers and elastomers, ozone attack on stretched rubber, deformation of precrystallized rubber at low temperature, and location of the fracture phase in thermoplastic elastomers.

D. Limitations of the ESR Method

In spite of the rather extensive use made of ESR in recent years as a fracture analysis tool, and the number of satisfying results claimed by its proponents, the technique does have certain limitations which tend to prevent full confidence in its measurements. It is worthwhile to list the most serious of these:

(1) Most polymeric systems do not suffer enough chain scission during fracture to permit ESR monitoring, so general statements about fracture in systems other than drawn nylon are tenuous. In some systems, bond rupture occurs only at a single small region, as discussed earlier. In other systems, fracture may occur largely as a result of large-segment rearrangements with scission playing only a peripheral role, and in still other systems scission may not occur at all.

(2) Heterolysis, rather than homolysis, may occur or even dominate during scission. These events would not produce ESR-detectable radicals. Ozone attack in rubber is generally considered heterolytic, for instance, although radicals are observed as well, presumably via secondary reactions.

(3) As mentioned earlier, it is common to correct for the ESR radical count for decay by employing appropriately determined kinetic models. However, it cannot be guaranteed that certain decay mechanisms are not operative, because they may be too rapid to be observed by the spectrometer. Radical migration to the stable side-chain position is an example of an extremely rapid radical reaction. If such reactions are involved in radical decay, the ESR spectrometer would underestimate the extent of bond scission.

(4) Another form of secondary radical reaction that would lead to underestimation of scission involves not radical decay but rather a radical-induced

rupture of an adjoining chain without the creation of a new radical. Using a main-chain rupture in a polyethylene sequence for illustration, we see:

$$
\begin{array}{cc}
\text{H} & \text{H} \\
| & | \\
\sim\!\!\sim\!\!\sim\text{C}\!\!-\!\!\text{C}\!\!\sim\!\!\sim\!\!\sim \\
| & | \\
\text{H} & \text{H}
\end{array}
\longrightarrow
\begin{array}{cc}
\text{H} & \text{H} \\
| & | \\
\sim\!\!\sim\!\!\sim\text{C}\cdot + \cdot\text{C}\!\!\sim\!\!\sim\!\!\sim \\
| & | \\
\text{H} & \text{H}
\end{array}
$$

The active radical may now abstract a proton form an adjoining chain:

$$
\begin{array}{c}
\text{H} \\
| \\
\sim\!\!\sim\!\!\sim\text{C}\cdot \\
| \\
\text{H} \quad \text{H} \quad \text{H} \\
\qquad\quad | \quad | \\
+ \sim\!\!\sim\!\!\sim\text{C}\!\!-\!\!\text{C}\!\!\sim\!\!\sim\!\!\sim \\
\qquad\quad | \quad | \\
\qquad\quad \text{H} \quad \text{H}
\end{array}
\longrightarrow
\begin{array}{c}
\text{H} \\
| \\
\sim\!\!\sim\!\!\sim\text{C}\!\!-\!\!\text{H} \\
| \\
\text{H} \\
\\
\qquad\qquad\quad \text{H} \\
\qquad\qquad\quad | \\
+ \sim\!\!\sim\!\!\sim\overset{\cdot}{\text{C}}\!\!-\!\!\text{C}\!\!\sim\!\!\sim\!\!\sim \\
\qquad\qquad\quad | \quad | \\
\qquad\qquad\quad \text{H} \quad \text{H}
\end{array}
$$

The presence of the side-chain radical weakens the main-chain bonds once removed from the radical, since upon rupture one of the newly unpaired electrons may enter into a π bond with the radical:

$$
\begin{array}{ccc}
& \text{H} & \text{H} \\
& | & | \\
\sim\!\!\sim\!\!\sim\overset{\cdot}{\text{C}}\!\!-\!\!\text{C}\!\!-\!\!\text{C}\!\!\sim\!\!\sim\!\!\sim \\
| & | & | \\
\text{H} & \text{H} & \text{H}
\end{array}
\longrightarrow
\begin{array}{ccc}
\text{H} & \text{H} \\
| & | \\
\sim\!\!\sim\!\!\sim\text{C} = \text{C} + \cdot\text{C}\!\!\sim\!\!\sim\!\!\sim \\
| & | & | \\
\text{H} & \text{H} & \text{H}
\end{array}
$$

The above rupture would be expected due to the weakening of the bond. An ESR spectrometer would see only one scission by counting the radicals formed to this point, while two chains would have been ruptured. The process can continue, with the result being that the spectrometer may underestimate the scissions by orders of magnitude. In fact, Zhurkov [10] has argued that this is in fact the case. Such a conclusion, if thoroughly verified, would clearly undermine the entire applicability of the ESR method.

(5) Even if the above problems are circumvented, ESR is deficient in that it has no ability to tell just where in the material the scissions are occurring. Although it has been possible to place the nylon fracture radical in the amorphous rather than the crystalline portion of the fiber structure, ESR cannot tell whether bonds are being broken at every tie-chain containing region or at relatively more dispersed positions, perhaps at the defects associated with microfibril ends. This is a distinction of considerable importance to the proper modeling of the fracture physics.

III. ASSOCIATED STUDIES

Given the above-mentioned ambiguities in the ESR results, it is not surprising that in recent years attention has turned to associated analytical techniques that can corroborate or modify the fracture models derived from ESR data. Some of the more active of these studies will be mentioned briefly here. The reader is also directed to chapters in this volume dealing with stress-active infrared absorption spectroscopy and mass spectroscopy.

A. Gel Permeation Chromatography

The ESR measurement of approximately 10^{18} scissions/gm at high fracture is high enough to expect that a measurable decrease in molecular weight may accompany fracture in nylon fibers. To investigate this possibility, a specimen of nylon that had been loaded to fracture at room temperature was submitted to gel permeation chromatography (GPC) analysis using m-cresol at 95°C as the solvent. The molecular weight distribution was obtained from the elution curve by using the "Q factor" method in reference to a polystyrene calibration. Figure 5 shows the molecular weight distribution of the fractured specimen, as well as that of a virgin specimen that had received no load. A reduction in the high molecular weight component is apparent, and the number-average molecular weight as computed from these chromatograms drops from 34,800 to 30,000. This corresponds to $(6.02 \times 10^{23})[1/30,000 - (1/34,800)] = 3 \times 10^{18}$ scissions/gm, in rather good agreement with the ESR

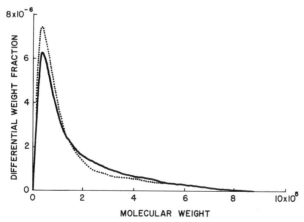

FIG 5. Effect of fracture on molecular weight distribution–GPC analysis. · · · ·, fractured; ———, virgin.

measurements. This value is also in agreement with viscosity molecular weight measurements by Crist [11]. Such tests lend credence to the ESR values, and provide some assurance that the more serious limitations to the ESR method described earlier are not operative in polyamides.

Although not as sensitive as ESR, GPC is not affected by secondary radical reactions and is thus more definitive. GPC also provides information as to which chains among the molecular weight distribution are broken during fracture, and this information may be of value in future model development.

B. Infrared Spectroscopy

As a result of secondary radical reactions following fracture, chemically distinct end groups may be formed at the scission site. In the earlier discussion of secondary reactions, CH_3 groups and unsaturated bonds were formed; other possibilities exist as well. Zhurkov [10] has reported success in monitoring the accumulation of such scission-induced groups by means of infrared spectroscopy, and some of his data show a large difference between broken bonds and observed radicals in polyethylene. IR shares GPC's advantage of monitoring stable entities rather than the marginally stable free radicals. Like GPC, IR is not as sensitive as ESR in monitoring fracture. However, the greatly enhanced sensitivity of the newly available Fourier Transform IR spectrometers promises to mitigate against this deficiency.

C. Small-Angle X-Ray Scattering

In another of his many pioneering investigations of fracture, Zhurkov [12] has reported success in using small-angle x-ray scattering (SAXS) to monitor the accumulation of microcracks developed during fracture, supposedly as a result of the coalescence of molecular scissions. Since SAXS is able in principle to measure the size, concentration, and spatial orientation of these microcracks, it offers promise of being able to unravel the problem mentioned earlier as to the spatial distribution of scission events. Zhurkov's data indicate that the microcrack concentration is much lower than the radical concentration, approximately 10^{15} cm^{-3}. This number is approximately equal to the number of microfibril ends in the fiber structure, in support of Peterlin's contention that damage is associated with strain inhomogeneities in these regions. Such a finding tends to weaken the tie-chain contour length models, which states that scission is predicted more uniformly throughout the specimen. The correct balance between these two viewpoints is currently a matter of some controversy, and work is now proceeding aimed at its resolution.

ACKNOWLEDGMENTS

This manuscript was prepared under the sponsorship of the U.S. Army Research Office, Contract DAAG29-76-0044. The author also thanks Prof. K. L. DeVries for his many helpful suggestions.

REFERENCES

1. A. S. Krausz and H. Eyring, "Deformation Kinetics." Wiley, New York, 1975.
2. H. H. Kausch, *J. Polym. Sci., Part C* **32**, 1–44 (1971).
3. S. N. Zhurkov, *Int. J. Fract. Mech.* **1**, 311–323 (1965).
4. C. P. Poole, "Electron Spin Resonance." Wiley, New York, 1976.
5. A. Peterlin, *Text. Res. J.* **42**, 20–30 (1972).
6. A. E. Brotskii, *Dokl. Akad. Nauk* **156**, 1147–1156 (1967).
7. K. L. DeVries, B. A. Lloyd, and M. L. Williams, *J. Appl. Phys.* **42**, 4644–4653 (1971).
8. B. A. Lloyd, Ph.D. thesis, Univ. of Utah, Salt Lake City, Utah, 1972.
9. K. L. DeVries and D. K. Roylance, *Progr. Solid State Chem.* **8**, 283–335 (1973).
10. S. N. Zhurkov, V. A. Zakrevskii, V. E. Korsukov, and V. S. Kuksenko, *Sov. Phys.— Solid State* **13**, 1680–1688 (1972).
11. B. Crist, *J. Polym. Sci., Part A-2* **16**, 485–500 (1978).
12. S. N. Zhurkov, V. S. Kusenko, and A. I. Slutsker, "Fracture 1969," pp. 531–544. Chapman & Hall, London, 1969.
13. D. K. Roylance, Ph.D. thesis, Univ. of Utah, Salt Lake City, Utah, 1968.

14

Stress Mass Spectrometry of Nylon 66*

MICHAEL A. GRAYSON and CLARENCE J. WOLF

MCDONNELL DOUGLAS RESEARCH LABORATORIES
MCDONNELL DOUGLAS CORPORATION
ST. LOUIS, MISSOURI

I. INTRODUCTION

The mechanical degradation of polymeric materials has emerged as a fundamental problem of major industrial importance. Of particular concern is the use of these materials in structural applications when little is known about their long-term stability. Considerable interest in this area exists, and a new field called mechanochemistry has developed (Casale *et al.* published an excellent review of this field [1]). Basically, mechanochemistry involves the study of chemical reactions for which the activation energy is supplied mechanically.

* This research was conducted under the McDonnell Douglas Independent Research and Development Program.

Stress-induced degradation or fracture of polymeric materials is a complex process in which a sequence of partially understood events occurs on both the molecular and macroscopic levels. Taylor [2] proposed a theory for fracture of inorganic glasses in which the time to failure is expressed as a function of applied stress. Recently, Zhurkov [3,4] applied these concepts to polymer glasses to develop a kinetic theory of fracture. Andrews [5] provided a concise review of the kinetic theory of fracture in polymer glasses. This theory assumes that primary bond cleavage is the dominant process leading to macroscopic failure of the polymer. A modified Arrhenius relation was developed which describes the rate of bond rupture as a function of energy, stress, and temperature:

$$R = R_0 \exp[-(U_{ab} - \beta\sigma)/kT], \tag{1}$$

where R is the rate of bond rupture in the polymer chain, R_0 is the fundamental frequency factor, U_{ab} is the activation energy for bond rupture, β is a constant characteristic of the polymer, σ is the applied stress, k is the Boltzmann constant, and T is the temperature (K). At zero applied stress, Eq. (1) reduces to the classical Arrhenius equation relating the rate of bond rupture to the temperature. According to the kinetic theory of fracture, a close correlation should exist between mechanical degradation and thermal degradation. The elementary events in both processes are the cleavage of chemical bonds in the polymer chain accompanied by the formation of free radicals on the ruptured ends of the chain. The primary free radicals rapidly react and/or rearrange to form more stable radicals which may lead to the generation of products which evolve during mechanical degradation. A close correlation between the volatile products observed during mechanical and thermal degradation strongly suggests that both the primary and secondary reactions following degradation are the same.

Stress mass spectrometry (stress MS) is a relatively new application of mass spectrometry to study the mechanical degradation of polymeric materials [6–8]. Materials are subjected to a stress, either mechanical or thermal, and the resultant products are analyzed mass spectrometrically. The entire experiment, including application of stress, is performed directly in the ion source housing of the mass spectrometer. A time-of-flight mass spectrometer (TOFMS) is ideally suited for these studies because it has a large, open ion source and produces 10,000 mass spectra per second. The samples can be stressed, fractured in tension, or abraded (sawed) while the mass spectra are recorded.

A. Background

Several investigators have conclusively shown that bond rupture occurs when stressed polymers fail. Free radicals have been detected in a variety of polymers under stress by electron spin resonance (ESR) spectroscopy. Polyethylene [3,4,9], polypropylene [9,10], polyethylene terphthalate [3,11], aliphatic polyamides [3,4,9,11–13], natural silk [3] and styrene–butadiene rubber [14] all form free radicals when mechanically stressed. Regel *et al.* [15–17] investigated the products formed during the mechanical degradation of a series of polymers and compared them with the products formed during thermal degradation. They concluded that the volatile products released from these two forms of degradation were the same and furthermore stated "that the mechanical failure of polymers may be regarded as stress-activated thermal degradation." [16] Baumgartner *et al.* [18] investigated the mechanical degradation of filled and unfilled elastomers. They concluded that "the mechanisms for thermally and mechanically induced decomposition appear to be equivalent." [18]

Most of the products observed by Regel *et al.* [15,17] as well as by Baumgartner and coworkers [18] are similar to those that are expected from incomplete polymerization. Several investigators have recognized the importance of this problem and have attempted to remove residual monomer and other low-molecular-weight compounds by thermal annealing of the specimens in vacuum. For example, Amelin *et al.* [16] stated that "volatile products which might have been in the specimens before the experiment began were removed by vacuum aging at 100–150°C." However, little analytic work was performed to determine the identity of indigenous volatile compounds or the success of purification procedures employed in the various studies. To resolve the ambiguities surrounding the origin of low molecular weight products of mechanical degradation, analytic methods for characterization of indigenous volatiles in polymers were employed by Grayson *et al.* [6] in a recent study of mechanical degradation in polystyrene. They concluded that most, if not all, of the styrene monomer observed during stress and/or failure of this material is due to unreacted styrene monomer trapped within the polymer matrix. Two methods were used to remove the indigenous styrene from the sample: vacuum annealing and fractional reprecipitation. Only the latter method, which was not used by the previous investigators, was effective in removing indigenous styrene monomer from the polymer matrix. No styrene monomer was detected during stress MS of the fractionally reprecipitated polystyrene. Based on the known detection limit of the apparatus ($\sim 10^{-10}$ gm/sec), the surface density of radicals at the fracture surface is less than 10^{10}/mm^2, in excellent agreement with the results of Nielsen *et al.* [19] who studied the fracture process in polystyrene by small-angle x-ray scattering and ESR spectroscopy.

The work of Grayson and coworkers [6] has important implications concerning stress mass spectrometry studies of polymers formed by free-radical polymerization. Since these polymers are difficult to free from traces of unreacted monomer and the monomer is expected to be a major mechanical degration product, stress mass spectrometry studies of such polymers cannot provide conclusive evidence of stress-induced degradation.

It is thus desirable to investigate another polymer which is free of this problem. We chose a condensation polymer, nylon 66, as a candidate for further stress MS studies.

Considerable work has been reported on the mechanical degradation of nylon 66 and its related compound, nylon 6. For clarity, the structures of these two compounds are:

$$\text{Nylon 66} \quad \left[\text{C} \overset{\text{O}}{\diagup} (\text{CH}_2)_4 \text{C} \overset{\text{O}}{\diagup} -\text{NH}-(\text{CH}_2)_6\text{NH} \right]_n$$

and

$$\text{Nylon 6} \quad \left[\text{NH}-\text{C} \overset{\text{O}}{\diagup} -(\text{CH}_2)_5 \right]_n$$

The chemical structures of the two polymers are similar. The free radicals produced by mechanical degradation of both these materials have been extensively investigated by ESR spectroscopy [9,11,12,20–23]. Apparently three primary radicals are formed during the mechanical degradation of either nylon at temperatures above $130°K$:

$$\text{~~~~CH}_2-\overset{\cdot}{\text{C}}\text{H}_2, \quad \cdot\text{CH}_2-\text{NH ~~~~} \quad \text{and} \quad \cdot\text{CH}_2\text{C} \overset{\text{O}}{\diagup} -\text{NH ~~~~}$$

These primary radicals rapidly undergo hydrogen abstraction to form the secondary radical.

$$\text{~~~~CH}_2\text{C} \overset{\text{O}}{\diagup} -\overset{\cdot}{\text{N}}-\text{CH}_2\text{~~~~}$$

which is stable at higher temperatures and has been developed several times [11,20]. The free-radical concentration following degradation is large, 10^{17} and 5×10^{17} radicals$\cdot$cm^{-3} for nylon 66 and nylon 6, respectively [23]. Backman and DeVries [9] reported that the surface concentration of free radicals in fractured nylon 66 is about 10^{13} radicals$\cdot$cm^{-2} so that the free radicals must be distributed throughout the ruptured sample and not restricted to the surface.

It is interesting to note that the total number of polymeric chains passing through a plane perpendicular to the stretch direction is about 10^{15} cm^{-2} [12]. Since total radical concentration is 10^{17} cm^{-3}, the chain rupture process must occur throughout the sample and not only at the point of catastrophic failure.

II. EXPERIMENTAL TECHNIQUE

In this work specific experiments were performed to: (a) characterize the indigenous volatile compounds in the nylon 66 samples; (b) detect and identify the major products of mechanical degradation; (c) detect and identify the major products of thermal degradation; and (d) characterize the indigenous volatile compounds in stressed samples.

A. Sample Characterization

Virgin nylon 66, free of additives, was obtained from the E. I. du Pont de Nemours and Company in the form of a drawn monofilament 0.2 mm in diameter. The indigenous volatile compounds were determined with the aid of a vaporization-gas-chromatography/mass-spectrometry (VGC/MS) and evolved-gas-analysis/mass-spectrometry (EGA/MS). VGC/MS was specifically developed to determine trace quantities of organic compounds trapped on or in Lunar fines and carbonaceous chondrite meteorites [24–27]. A schematic representation of this technique is given in Fig. 1. The sample (typically a few milligrams) is loaded into a precleaned quartz tube. After the tube is flushed with helium for several minutes, the isolation valve is opened, and the tube with sample is inserted into the hot zone. The U bend in the GC column is simultaneously immersed in a liquid nitrogen bath. The compounds evolved from the heated sample are transported by the gas flow into the column where they are trapped at the U bend. A second He carrier gas supply is connected immediately behind the isolation valve and flushes

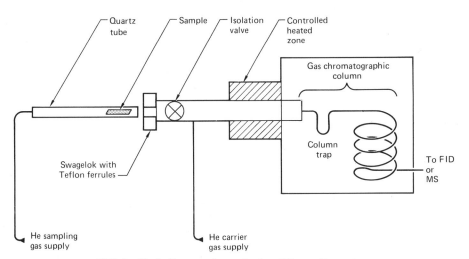

FIG. 1   Block diagram of vaporization GC sampling system.

the outside of the sampling tube to prevent back diffusion of the vapors. The sample is heated to the desired temperature (usually 200°C) for the desired length of time (typically 15 min). All the volatile compounds evolved during this heating interval are trapped directly on the GC column. When the heating period is complete, the sample tube is removed from the hot zone and the isolation valve is closed. The liquid nitrogen bath is then removed from the U bend, and the chromatographic analysis is initiated.

The outlet of the GC column was interfaced [28] to a mass spectrometer (AEI MS-30). Fast-scan mass spectra of the events were continuously recorded in computer-compatible format during the chromatographic analysis. The mass spectral data were subsequently interpreted with the aid of a computer facility [29].

The VGC/MS characterization technique is well suited for the analysis of volatile compounds provided the compounds of interest have sufficient vapor pressure to be analyzed chromatographically. Since EGA/MS has a higher sensitivity for compounds of low volatility, it was used to characterize the nylon 66 for compounds of low vapor pressure. The sample (several tenths of a milligram) was placed in the solids-probe sampling cup, inserted into the ion source, and temperature programmed at a linear rate from 50–400°C. During the temperature program, the mass spectrometer was continuously scanned, and data were recorded in digital format for later interpretation.

B. Mechanical Deformation Experiments

A time-of-flight mass spectrometer (Bendix model 12–101) was used for the stress MS experiments. A special apparatus (Fig. 2) was designed, constructed, and mounted to the ion source housing flange of the TOFMS to permit tension loading of the pure nylon 66 monofilament. The apparatus consists of a split cylindrical mandrel around which the sample is wrapped. One-half of the mandrel is attached to a fixed lintel mounted on the ion source flange, while the other half is attached to a loading rod which passes through a vacuum feedthrough in the flange. Mechanical loads are applied to the sample in two ways: static and impulsive. The static load is applied by adding known weights to the carriage attached to the loading rod. Impulsive loads are applied by dropping a sliding weight, which is concentric with the loading rod, against a stop. Impulsive loading is required so that sample failure is forced to occur in a narrow time span (a few seconds) during which the mass spectral data acquisition system records data. A known weight is added (static mode) and an impulsive load is applied. If the sample does not fail in several seconds, the sequence is repeated until sample failure. Extension of the sample is measured by a dial gauge attached to the loading rod which is referenced against the bottom of the source flange.

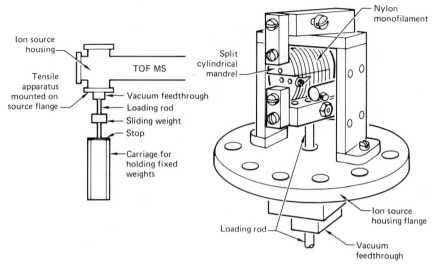

FIG. 2   Apparatus for stressing monofilaments in the ion source housing of a time-of-flight mass spectrometer.

C. Thermal Degradation

Thin films of nylon 66 were thermally degraded directly inside the ion source housing of the TOFMS. A filament pyrolyzer was inserted via a vacuum lock without disturbing the mass spectrometer vacuum system. The platinum filament (0.25 mm diameter) is resistively heated by a capacitor discharge circuit which ensures rapid heating of the filament to the desired temperature within several milliseconds [30]. A low-mass chromel–alumel thermocouple was tack welded to the filament to calibrate the temperature of the filament as a function of the charging voltage on the capacitor. The thermocouple was removed before thermal degradation experiments were conducted. This system provides a means of rapidly and accurately heating thin films from 100–900°C. The platinum filament is cleaned by heating in air several times prior to the next degradation.

D. Data Handling

Mass spectral data from both the mechanical and thermal degradation of nylon 66 were obtained with the rapid events mass spectral data acquisition (REMSDA) system. This system was specifically designed to rapidly record data immediately prior to, during, and after the sample is either mechanically or thermally stressed. The REMSDA system rasters z axis modulated mass spectra on an oscilloscope in a fashion similar to that described by Lincoln

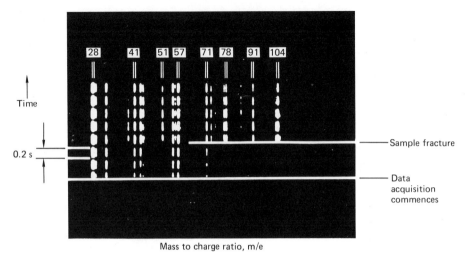

FIG. 3   Rapid events mass spectra data acquisition (REMSDA) display from the fracture of polystyrene.

[31]. The data are recorded by oscilloscope photography with a Polaroid camera. A typical REMSDA output is shown in Fig. 3. The mass-to-charge ratio of the ions is displayed along the x axis, elapsed time along the y axis, and peak intensity along the z axis (perpendicular to the page). The REMSDA system is triggered to record data prior to the occurrence of the stress event, thereby providing a background spectrum. A signal driven by the mass spectrometer clock is used to add steps to the ramp voltage which drives the time axis, thereby recording time on the REMSDA output.

The detection limit of the TOFMS–REMSDA system was determined by adding known aliquots of vapors at controlled rates. The minimum quantity of gas which yields a readable line on the REMSDA photographic display is approximately 10^{-9} gm/sec.

III. RESULTS AND DISCUSSION

The major indigenous volatile compounds in virgin nylon 66 were determined by EGA/MS to be water and cyclohexamethylene adipamide. The minor volatile compounds were characterized by VGC/MS. Approximately 70 mg of nylon 66 was heated to 210°C for 15 min while the evolved compounds were trapped on the chromatographic column. The chromatographic parameters shown in Table I were used to obtain the chromatogram shown in Fig. 4. The identified compounds were present in concentrations less than 10^{-6} wt% and are most likely contaminants attributable to the surface

TABLE I

*Sampling and Chromatographic Conditions
for Characterization Experiments*

Hot zone temperature	----------------------	$210^{o}C$
Sampling period	--------------------------	15 min
Sampling gas supply	------------------------	5-10 cm^3 atm/min
Carrier gas supply	--------------------------	22 cm^3 atm/min
Column:	Length ----------------------	2 m (7 ft)
	Bore -------------------------	1.5 mm (1/16 in.)
	Support ----------------------	100/120 mesh Suplecoport
	Phase ------------------------	SP 2000
	Loading ----------------------	10%
Temperature program:		

2 min at room temp

$15^{o}C$/min to $285^{o}C$

10 min at $285^{o}C$

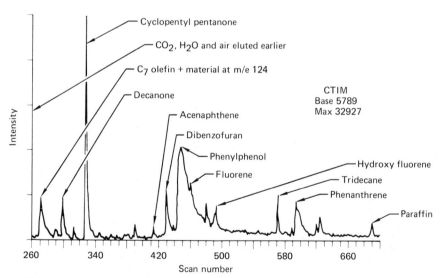

FIG. 4   Chromatogram of a VGC/MS analysis of the volatiles released from nylon 66 (heated for 15 min at 210°C).

finishes and die lubricants applied during manufacture. When the same sample was reheated, only a trace of p-phenylphenol remained.

The mechanical degradation experiments were performed with 2-m long sections of the monofilament. This length accommodated approximately 20 turns of monofilament around the mandrel. Only the ends of the monofilament were attached to the mandrel; thus, a single break in the monofilament terminated the experiment. Single breaks were rare, but when they occurred they were close to the filament attachment points. Typically, the monofilament broke in two or three places; in one instance ten breaks were observed.

The sample loading sequence utilized a 12-kg static load followed by an impulsive load of 0.5 kg dropped from a height of 15 cm after the addition of each static load. REMSDA photographs were recorded during the application of the impulsive load. The sample extension prior to failure was typically 10%. No evolved compounds were detected during loading regardless of the extension or load on the sample. Two distinct types of fracture were observed during these experiments: one led to a clean sharp break and the other led to a frayed or fibrillated tip. These fracture types are subsequently referred to as clean and fibrillated breaks. (See Fig. 5 for a photograph of typical fracture surface tips.) Clean breaks are characterized by evidence of flow and necking at the fracture surface. Deformation of the sample along its axis was limited to several monofilament diameters. Fibrillated breaks, however, were frayed into several strands, and the sample had the appearance of exploding. In this case, deformation of the sample along its axis was on the order of 10 to 20 monofilament diameters.

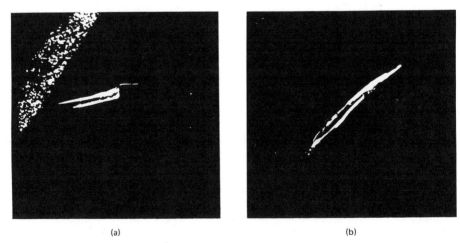

(a)   (b)

FIG. 5   Typical fracture surfaces from failure of nylon 66 monofilament: (a) clean fracture surface; (b) fibrillated fracture surface.

It is particularly noteworthy that whenever one or more of the fracture surface tips was fibrillated, volatile compounds were detected by stress mass spectrometry. This relationship was consistently observed throughout our studies of nylon 66.

A typical REMSDA (70 eV ionization potential) display of the data obtained during failure of the monofilament is shown in Fig. 6. The mass range displayed is from m/e 13 to $\sim m/e$ 140. After data acquisition commences, the background in the mass spectrometer is recorded for 0.3 sec, at which time sample failure occurs. Within 0.5 sec the majority of the volatile degradation products are evolved, and within 1.0 sec after failure the concentration in the ion source reduces to that of the background. The background ions are due to water (17,18), air (28,32), hydrocarbon contaminants in the mass spectrometer (41,43,55,57,69,71), and doubly ionized mercury (99–101). Upon sample fracture, a complex mixture of ions with m/e less than 100 is evolved. The mass spectral data shown in Fig. 6 represent the time-resolved mass spectrum of that mixture.

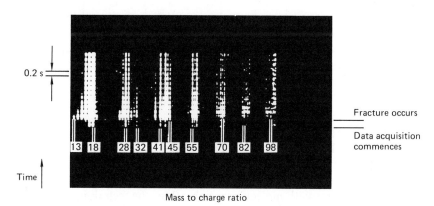

FIG. 6   REMSDA display of mass spectra of compounds evolved upon fracture of a nylon 66 monofilament.

The data are summarized in Table II where the ions observed upon fracture are tabulated according to relative concentration and listed as small (S) medium (M), or large (L). It is difficult to positively identify compounds from the mass spectrum of a mixture of compounds. Therefore, the stress MS experiments were repeated over a range of low ionization potentials varying from 9–15 eV to permit identification of some of the compounds evolved upon fracture. Low ionization potentials simplify the mass spectral fragmentation (i.e., the cracking pattern) by enhancing the relative intensity of the parent ion. In some instances, when the ionization potentials of the compounds in question are in the proper range, they can be used for identification.

TABLE II

Ions Observed from the Mechanical and Thermal Degradation of Nylon 66

m/e^a	Mechanical degradation	Thermal degradation [b]	Background [c]	m/e^a	Mechanical degradation	Thermal degradation [b]	Background [c]
13	S			44	L	L	
14	S			45	L	S	
15	M			55	L	L	X
16	L			56	L		
17	L		X	57	L		X
18	L	L	X	58	S	M	
19	L			67	M		
26	S			68	M		
27	M	L		69	M		X
28	L	L	X	70	M		
29	L	L		71	M		X
30	S	M		81	S		
31	S			82	S		
32	L	M	X	83	S		
39	L	L		84	M	L	
41	L	L	X	95	M		
42	L	L		96	M		
43	L	L	X	98	S		

[a] Ionization potential: 70 eV.
[b] Decomposition at 400°C.
[c] Ions present in the background.

For example, the ion at m/e 28 can be due to N_2 (I.P. = 15.58 eV), CO (I.P. = 14.01 eV), or C_2H_4 (I.P. = 10.5 eV) [32]. Stress MS experiments with an ionization potential of 11.0 eV indicated that at least part of the m/e 28 line is due to ethylene. Unfortunately, the number of compounds possible at a given mass increases with molecular weight; thus, the technique has limited utility. Nevertheless, on the basis of such experiments, positive identification was obtained for the evolution of ammonia and ethylene, and tentative identifications were obtained for ethane, butenol, and butenal. In addition, both CO_2 and H_2O are probably evolved, although the absolute determination of these compounds is complicated by their presence in the background of the mass spectrometer. Another compound besides CO_2 whose ionization potential is less than 11.5 eV (possibly propane; I.P. = 11.07 eV) was observed at m/e 44. Cyclopentanone was also identified from its characteristic fragment ions at m/e 84, 55, 39, 42, and 56. These compounds, i.e., NH_3, C_2H_4, CO_2, C_2H_6, C_4H_7OH, C_3H_5CHO, and cyclopentanone, were not detected during the characterization studies of the monofilament by VGC/MS and EGA/MS and thus are interpreted as mechanical degradation products.

It is important to compare the results of mechanical degradation experiments with those from thermal degradation. However, a similar system employing the same general techniques and data handling procedures is required to minimize instrumental and/or experimental artifacts. Therefore, the platinum resistance pyrolyzer was used to thermally degrade thin films of nylon 66. The films were deposited from a benzene (5%) phenol solution. The pyrolysis process occurs in the ion source housing of the TOFMS, and the data were acquired with the REMSDA system. Thus all of the experimental conditions were nearly identical in the thermal and mechanical experiments.

A typical REMSDA output from the thermal degradation of approximately 10 ng of nylon 66 at 400°C is shown in Fig. 7. The major ions observed are listed in Table II together with those found in the background. The nylon 66 films were thermally degraded from 200° to 600°C. Spectra from a typical film heated to 200° and 600°C are shown in Fig. 8. In all experiments, the filament was heated to 150°C immediately prior to pyrolysis to remove the solvent. Nevertheless, a trace of phenol (m/e 94) appeared in the degradation spectra.

As expected, the spectra of the degradation products vary considerably with temperature. High temperatures favor the formation of low molecular weight species. Even at high temperatures, however, the thermal spectra are considerably simpler than the mechanical spectra. This may be an artifact of the experimental system in that all of the mechanical degradation products are released in a few milliseconds after fracture and produce a high concentration of compounds in the ion source over a short time interval.

The most abundant compound produced during thermal degradation is cyclopentanone. Several other typical hydrocarbon ions (m/e 27, 29, 41, and

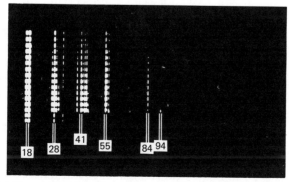

FIG. 7   Volatile products from thermal degradation of a thin film of nylon 66 in the ion source housing of a TOFMS (70 eV, 400°C).

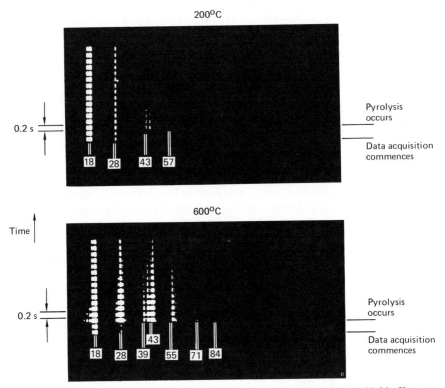

FIG. 8   Comparison of thermal degradation at 200°C and 600°C of nylon 66 thin films.

43) were also observed. Ammonia was positively identified by ionization potential studies, but apparently ethylene is not present. It is difficult to positively state that either CO_2 or H_2O is present, but the results suggest that both are present.

Considerable work on the thermal degradation of nylon 66 has appeared and is briefly summarized by Peebles and Huffman [33]. The thermal degradation products fall into three broad classes: (a) highly volatile, low molecular weight compounds; (b) less volatile, higher molecular weight compounds, which generally condense just outside of the hot zone; and (c) a polymeric residue. The highly volatile material, of primary interest in our studies, had been examined by Achhammer et al. [34] and Kamerbeek et al. [35]. According to Achhammer, the volatile degradation products from the decomposition of nylon 66, in order of decreasing yield, are CO_2, H_2O, cyclopentanone, C_1 to C_6 hydrocarbons, and CO. Kamerbeek et al. investigated the low-temperature degradation (305°C) and conclusively proved that ammonia is a major degradation product. In fact, they reported that more than half of the nitrogen in the polymer is released in the form of ammonia after heating at 305°C for 100 hr. The other major volatile products are H_2O and CO_2. They postulate that chain-end groups are formed by the hydrolysis of the amide groups in the polymer chain, i.e.,

$$\sim\!\sim\!\sim NH\!-\!\overset{\displaystyle O}{C}\!\sim\!\sim\!\sim + H_2O \Longrightarrow \sim\!\sim\!\sim NH_2 + HO\overset{\displaystyle O}{C}\!\sim\!\sim\!\sim$$

to form an amine and acid chain end. Water is produced from the rearrangement of a carbamide radical to H_2O and a nitrile. They further speculate that the small molecules, ammonia, carbon dioxide and water, are produced by the bimolecular reaction of the chain-end groups:

$$\sim\!\sim\!\sim\overset{\displaystyle O}{C}\!-\!NH\!-\!(CH_2)_6NH_2 + NH_2(CH_2)_6NH\overset{\displaystyle O}{C}\!\sim\!\sim\!\sim \Longrightarrow$$

$$NH_3 + \sim\!\sim\!\sim\overset{\displaystyle O}{C}\!-\!NH(CH_2)_6NH(CH_2)_6NH\overset{\displaystyle O}{C}\!\sim\!\sim\!\sim$$

$$\sim\!\sim\!\sim(CH_2)_6\overset{\displaystyle O}{C}\!-\!OH + HO\overset{\displaystyle O}{C}\!-\!(CH_2)_6 \Longrightarrow H_2O + CO_2 + \sim\!\sim\!\sim(CH_2)_6\overset{\displaystyle O}{C}\!-\!(CH_2)_6\!\sim\!\sim\!\sim$$

$$\rangle NH + HO\overset{\displaystyle O}{C}\!-\!(CH_2)_4\!\sim\!\sim\!\sim \Longrightarrow H_2O + \rangle N\!-\!\overset{\displaystyle O}{C}\!-\!(CH_2)_4\!\sim\!\sim\!\sim.$$

It is difficult to directly correlate mechanical and thermal degradation. A "standard degradation process' does not exist. The effect of final temperature, film thickness, heating rate (temperature rise time) and surrounding atmosphere all affect the thermal degradation product distribution. The problems associated with obtaining reproducible thermal pyrolysis data were recently reviewed [36].

General agreement exists between the products observed in mechanical and thermal degradation. In both cases, cyclopentanone, ammonia, carbon dioxide, and water are observed. It is important to note, however, that major differences, particularly in the product distribution, exist. For example, the most pronounced thermal degradation product is cyclopentanone. While this compound is formed during mechanical degradation, many other hydrocarbon-type ions, including ethylene, are observed in the spectra. The ammonia-to-cyclopentanone ratio is larger in the mechanical degradation experiments. At present, the reasons for these differences are not known, although it must be noted that the apparent equilibrium temperature surrounding the mechano-radicals is considerably less than the temperature attained in thermal degradation.

The mechanisms which have been postulated by Kamerbeek [35] to account for ammonia, water, and carbon dioxide product in thermally degraded nylon do not appear as viable mechanical degradation processes. The mechano-reactions occur at or near the relatively cool fracture surface within a few milliseconds of fracture, conditions which do not favor the complex sequence of bimolecular chain-end reactions proposed by Kamerbeek. Although the products of mechanical and thermal degradation are essentially the same, different mechanisms may be operative.

Our previous studies of polystyrene [6] failed to reveal any correlation between the products of mechanical and thermal degradation. In fact, we showed that the major product reported by Regel and coworkers [15–17] in the mechanical degradation of polystyrene, i.e., styrene, was not a degradation product but indigenous monomer. The similarity in products from thermally and mechanically degraded nylon 66 suggests some correlation in these processes.

As indicated earlier, mechanical degradation products are observed only when the fracture surface is fibrillated. Two alternate explanations for this observation can be offered: (1) although stress-induced chemical reactions always occur, the products are trapped in the matrix and are released only when a large surface area of the matrix volume (i.e., the fibrillated fracture) is exposed to the ion housing vacuum; or (2) a unique fracture process occurs concomitant with fibrillation to yield a large fraction of volatile compounds. Backman and DeVries [9] studied the fracture of nylon 66 and concluded that the free radicals observed by ESR spectrometry were not restricted to the surface but were distributed throughout the sample. Campbell and Peterlin [12] reached an identical conclusion in their studies of nylon 66. Therefore, we utilized the highly sensitive technique of VGC to measure and compare the concentration of organic compounds released from a stressed and an unstressed nylon sample. VGC has been successfully utilized to measure sub parts per million of organic compounds in meteorites [26,27], and the sensitivity of the system as used in these experiments was approximately 10^{-10} gm/sec. The samples studied were subjected to identical environments from the time they were cut from the spool to the time of analysis, except that one of the monofilaments was subjected to a mechanical load. The VGC chromatograms of a stressed and unstressed sample are shown in Fig. 9. They are almost identical, indicating that the first hypothesis, i.e., trapped products are released because of the large surface area exposed during fibrillation, is not true. Thus we conclude that fibrillation and the associated volatile product formation result from the same process.

DeVries et al. [22] found that nylon 66 fractured in torsion (shear) produced a highly frayed structure "splitting the monofilament into many smaller fibers." They also found that the free-radical concentration yield from nylon 66 fractured in tension was some 10–100 times greater than the corresponding yield produced in torsion. They suggested that this difference may be because in tension, the fracture energy breaks primary (main-chain)

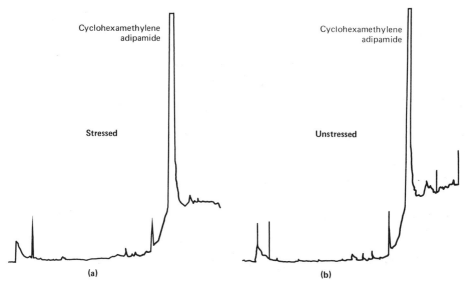

FIG. 9   Comparison of volatiles released from stressed and unstressed nylon 66.

bonds while in torsion, secondary (van der Waals) bonds are broken. Nevertheless we, as have other investigators [3,4,15–17], have assumed that volatile products and broken main-chain bonds, i.e., free-radical, are related. Quite possibly another explanation, such as unimolecular decompositions, may be required. Further work is needed before additional speculation on this topic is warranted.

IV. CONCLUSION

Stress MS has been applied to a study of the fracture process in nylon 66. Volatile products are observed only when the fracture surface is highly frayed or fibrillated. Several of the volatile products, cyclopentanone, NH_3, CO_2 and H_2O, are also thermal degradation products. Thus, some correlation exists between mechanical and thermal degradation, but the correspondence is not one-to-one and different mechanisms probably are involved.

ACKNOWLEDGMENTS

The authors acknowledge Dr. Ram Levy and Mr. Dale Fanter for their helpful discussions and suggestions and Mr. Eugene Arnold for his care and diligence in performing the thermal degradation experiments.

REFERENCES

1. A. Casale, R. S. Porter, and J. F. Johnson, *Rubber Chem. Tech.* **44**, 534 (1971).
2. N. W. Taylor, *J. Appl. Phys.* **18**, 943 (1947).
3. S. H. Zhurkov and E. E. Tomashevsky, *Conf. Proc.: Physical Basis of Yield and Fracture*, p. 200. Oxford Univ. Press, London and New York, 1966.
4. S. N. Zhurkov, V. A. Zakrevskyi, V. E. Korsukov, and V. S. Kuksenko, *J. Polym. Sci., Part A-2* **10**, 1509 (1972).
5. E. H. Andrews, "The Physics of Glassy Polymers" (R. N. Haward, ed.), p. 394. Wiley, New York, 1973.
6. M. A. Grayson, C. J. Wolf, R. L. Levy, and D. B. Miller, *J. Polym. Sci., Part A-2* **14**, 1601 (1976).
7. M. A. Grayson, R. L. Levy, D. L. Fanter, and C. J. Wolf, presented at the *24th Ann. Conf. Mass Spectrometry and Allied Topics*.
8. M. A. Grayson, R. L. Levy, C. J. Wolf, and D. L. Fanter, *Polym. Prepr. Amer. Chem. Soc., Div. Polym. Chem.* **17**(2), 764 (1976).
9. D. K. Backman and K. L. DeVries, *J. Polymer Sci Part A-1* **7**, 2125 (1969).
10. M. Sakaguchi, H. Yamakowa, and J. Sohma, *J. Polym. Sci., Part B* **12**, 193 (1974).
11. T. C. Chiang and J. P. Sibilia, *J. Polym. Sci., Part A-2* **10**, 2249 (1972).
12. D. Campbell and A. Peterlin, *J. Polym. Sci., Part B* **6**, 481 (1968).
13. S. N. Zhurkov, A. Ha. Savostin, and E. E. Tomashevsky, *Sov. Phys.—Dokl.* **9**, 986 (1965).
14. B. Eilis and J. F. Baugher, *J. Polym. Sci., Part A-2* **11**, 1461 (1973).
15. V. R. Regel, T. M. Muinov, and O. F. Pozdnyakov, *Conf. Proc.: Physical Basis of Yield and Fracture*, p. 194. Oxford Univ. Press, London and New York, 1966.
16. A. V. Amelin, T. M. Muinov, O. F. Pozdnyakov, and V. R. Regel, *Mekh. Polim.* **3**, 80 (1967).
17. A. V. Amelin, O. F. Pozdnyakov, V. R. Regel, and T. P. Sanfirova, *Sov. Phys.—Solid State* **12**, 2034 (1971).
18. W. E. Baumgartner, J. A. Hammond, G. E. Myers, and W. G. Stapleton, *Polym. Prepr., Amer. Chem. Soc., Div. Polym. Chem.* **14** (1), 448 (1973).
19. L. E. Nielsen, D. J. Dahm, P. A. Berger, V. S. Murtz, and J. L. Kardos, *J. Polym. Sci., Part A-2* **12**, 1239 (1974).
20. H. H. Kausch–Blecken Von Schmeling, *J. Macromol. Sci., Rev. Macromol. Chem.* **4**, 243 (1970).
21. K. L. DeVries and K. K. Roylance, and M. L. Williams, *J. Polym. Sci., Part A-1* **8**, 237 (1970).
22. K. L. DeVries, R. D. Luntz, and M. L. Williams, *J. Polym. Sci., Part B* **10**, 409 (1972).
23. A. Peterlin, "ESR Applications to Polymer Research" (P. Kinell, B. Ranby, and V. Runnström–Reio, eds.), p. 235. Wiley, New York, 1972.
24. R. L. Levy, C. J. Wolf, and J. Oro', *J. Chromatogr. Sci.* **8**, 524 (1970).
25. J. Oro', W. S. Updegrove, J. Gibert, J. McReynolds, E. Gil–Av, J. Ibaney, A. Zlatkis, D. A. Flory, R. L. Levy, and C. J. Wolf, *Proc. Apollo 11 Lunar Sci. Conf.* **2**, 1901. Pergamon, Oxford, 1970.
26. R. L. Levy, C. J. Wolf, M. A. Grayson, J. Gibert, E. Gelpi, W. S. Updegrove, A. Zlatkis, and J. Oro', *Nature* **227**, 1481 (1970).
27. R. L. Levy, M. A. Grayson, and C. J. Wolf, *Geochim. Cosmoschim. Acta* **37**, 467 (1973).
28. M. A. Grayson and R. L. Levy, *J. Chromatogr. Sci.* **9**, 687 (1971).
29. M. A. Grayson, J. M. Putnam, and F. J. Yaeger, *Int. J. Mass Spectrom. Ion Phys.* **22**, 365 (1976).
30. C. J. Wolf, R. L. Levy, and D. L. Fanter, *J. Fire Flammability* **5**, 76 (1974).
31. K. A. Lincoln, *J. Mass Spectrom. Ion Phys.* **2**, 75 (1969).

32. R. W. Kiser, "Introduction to Mass Spectrometry and its Applications," p. 308. Prentice–Hall, Englewood Cliffs, New Jersey, 1965.

33. L. H. Peebles, Jr., and M. W. Huffman, *J. Polymer Sci. Part A-1* **9**, 1807 (1971).

34. B. G. Achhammer, F. W. Reinhart, and G. M. Kline, *J. Res. Nat. Bur. Stand.* **46**, 391 (1951).

35. G. H. Kamerbeek, H. Kroes, and W. Grolle, *Soc. Chem. Ind. London, Monog.* **13**, 357 (1961).

36. C. J. Wolf and R. L. Levy, "Applications of the Newer Technique of Analysis" (I. L. Simmons and G. W. Ewing, eds.), p. 175. Plenum, New York, 1974.

15

Characterization of Polymer Decomposition Products by Electron Impact and Chemical Ionization Mass Spectrometry

D. A. CHATFIELD   *F. D. HILEMAN*
K. J. VOORHEES   *I. N. EINHORN*
          J. H. FUTRELL

FLAMMABILITY RESEARCH CENTER
DEPARTMENT OF MATERIALS SCIENCE AND ENGINEERING
UNIVERSITY OF UTAH
SALT LAKE CITY, UTAH

I. INTRODUCTION

The evaluation of combustion and thermolytic degradation products from both man-made and natural polymeric materials is an exceedingly difficult task. The variety of compounds produced can tax the most sophisticated separation techniques and elaborate detection methods. The unusual

241

toxicity of certain degradation products has sometimes required trace analyses on complex mixtures when only small amounts of sample are available [1].

For the past several years, we have been applying both electron impact (EI) and chemical ionization (CI) mass spectrometry to the task of compound identification. A mass spectrometric detector coupled to a gas chromatograph has been shown to be a powerful technique for the analysis of these complex mixtures. In this chapter, we wish to present several examples of how EI and CI mass spectrometric techniques have been used to analyze degradation products, identify component parts of rigid polyurethanes, and provide a method for determining the kinetics of the degradation of polyvinyl chloride.

II. MASS SPECTROMETRY

A. Electron Impact Techniques

The most common technique for generating a characteristic positive ion spectrum of a molecule is electron impact-mass spectrometry (EI-MS). Ion sources for conducting this kind of ionization in a reproducible way are very well developed [2]. This method is capable of both quantitative accuracy in a particular instrument and qualitative reproducibility of the characteristic mass spectrum on a daily basis in various mass spectrometer designs. Because of the high sensitivity achievable by EI-MS, determination of the molecular weight of compounds producing molecular ions as low as 1% are possible on microgram amounts of compound. Even in the event that a molecular ion is not present, large mass spectral data files have been developed to simplify the problem of compound identification [3].

Although the EI-MS technique is sensitive, means must be provided to separate the complex mixture of polymer degradation products prior to MS analysis. It is difficult to overemphasize the importance of fractionating these mixtures in a highly efficient way to ensure that mass spectra of single components, and not complex mixtures, are recorded at any given time. An especially valuable approach is the combination of a gas chromatograph-mass spectrometer (GC-MS) coupled to a computerized data acquisition package. The magnitude of the information content from the GC-MS analysis of volatile polymer thermolysis products has led to the necessity of storage of large data files. For example, Fig. 1 represents the flame ionization detector (FID) response from the pyrolysis of polyvinyl fluoride at 450°C. The compounds eluted over an 80-minute period and the entire mass spectrum from m/q 10 to 350 was scanned every 3.5 seconds. Thus, 1370 mass spectra were recorded, and 4.7×10^5 mass-intensity data were stored in

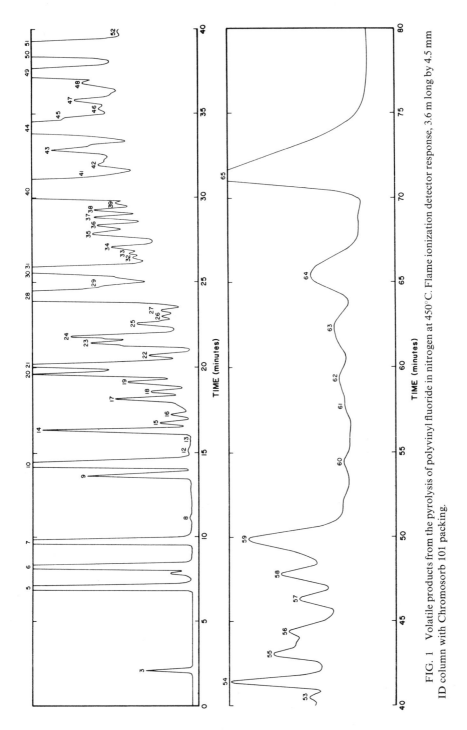

FIG. 1   Volatile products from the pyrolysis of polyvinyl fluoride in nitrogen at 450°C. Flame ionization detector response, 3.6 m long by 4.5 mm ID column with Chromosorb 101 packing.

243

these spectra. There were a minimum of 65 compounds in this mixture and MS techniques were used to identify 48 compounds. The MS spectra were useful in identifying several compounds that were not fully resolved by the GC (e.g., 30 and 31, 44 and 45, and 50 and 51) using selected ion monitoring techniques (to be described later). The chromatogram in Fig. 1 is representative of the magnitude of the analytical separation problem that is encountered in pyrolysis studies. Extending the resolving power of the GC by use of capillary columns for polymer degradation studies will be the topic of a future publication.

B. Chemical Ionization Techniques

A new approach to analytic mass spectrometry, in which the ion spectrum characteristic of a given molecule is generated in a series of chemical reactions, has been described in detail by F. H. Field and collaborators [4]. The particular class of chemical reactions giving rise to the characteristic spectrum is ion–molecule reactions between a plasma of ions formed from a reactant gas and a sample molecule. The technique involves admitting a high pressure (0.1 to 1.0 torr) of a reactant gas into the ion source of a specially modified mass spectrometer. The reactant gas is ionized via electron impact and the ions undergo several hundred collisions with neutral reactant gas molecules before they effuse from the ion source and enter the mass spectrometer. A reactant gas is chosen which produces a simple mass spectrum with reactive ions at low masses (below m/q 60). Methane, water, isobutane, and ammonia are the most commonly used positive ion reactant gases. Methane, for example, produces a plasma containing CH_5^+ (m/q 17), $C_2H_5^+$ (m/q 29), and $C_3H_5^+$ (m/q 41). The addition of compound M (at a partial pressure of less than 1% of the entire ion source pressure) generates characteristic CI spectra containing a protonated molecular ion, m/q M + 1, and often ion–molecule adducts at m/q M + 29 and M + 41.

Since CI spectra are developed via bimolecular ion–molecule reactions rather than through impact of an electron, there are a number of characteristics which distinguish it from EI spectra. Ion–molecule reactions which are observed under CI conditions are generally restricted to low energy processes, resulting in less fragmentation of the protonated molecular ion. If desired, all or most fragmentation can be suppressed by the choice of a suitable reactant gas, resulting in a single peak spectrum of the unknown compound.

C. Simultaneous EI–CI Mass Spectrometry

Both EI and CI mass spectrometry have important analytic applications and are highly complementary techniques. They have been combined in an

AEI MS-30 dual beam mass spectrometer for simultaneous EI and CI spectral recording of gas chromatographic effluents [5]. The GC-MS interface, shown in Fig. 2, splits the GC effluent into two parts with a silicone membrane separator. The fraction passing through the separator is transferred to an ion source where EI spectra are produced that are identical to conventional GC-MS spectra. The remaining effluent from the gas chromatograph is mixed with a reactant gas and bled into a separate high pressure ion source, where a characteristic CI spectrum is generated.

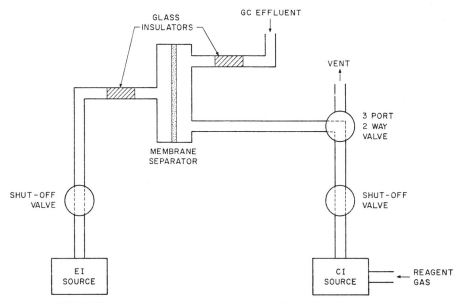

FIG. 2   Silicone membrane separator arrangement for dual ion beam operation.

The utility of simultaneous EI–CI spectra can be demonstrated by the examination of mass spectra from a compound that was encountered in routine GC-MS analysis. Figure 3 is a trace of the EI–CI mass spectrum that was recorded during the elution of a GC peak. The most intense peak in the methane–CI spectrum was m/q 60, suggesting that the molecule contains an odd number of nitrogen atoms and has a molecular weight of 59. The M − 1 ion in the CI spectrum (m/q 58) is characteristic of a primary rather than a secondary or tertiary amine, so the compound is propylamine. Since n-propyl amine shows a small m/q 59 peak and no m/q 58 peak under EI conditions, the EI spectrum completes the identification of the compound as isopropyl amine. The identity of this compound may be checked by comparison with EI data from mass spectral data files. EI spectra of amines are often not

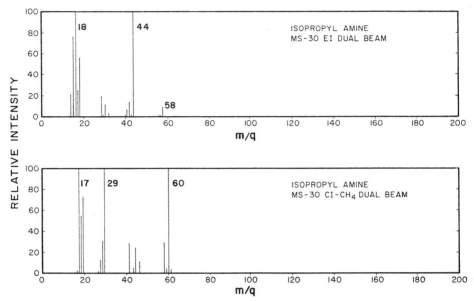

FIG. 3   Simultaneously generated electron impact and chemical ionization mass spectra of isopropylamine.

readily interpreted because of temperature stability, extensive fragmentation, and no molecular ion. However, amines pose no particular difficulties when handled by EI and CI-MS techniques.

Another example that demonstrates the utility of both EI and CI mass spectral data was provided by the analysis of the pyrolysis products of an aromatic polyimide. The EI spectra from a symmetrically shaped GC peak, shown in Fig. 4a, was not directly interpretable. The great abundance of m/q 93 and 103 and the 10 amu difference between them suggested that either the molecular ion was not present in the mass spectrum or that more than one compound was eluting from the GC column with nearly identical retention indices. The simultaneously generated CI spectrum, shown in Fig. 4b, was very helpful in resolving this problem. The ions at m/q 17, 19, 29, and 41 were generated by methane reactant gas (which contains a trace of water) under CI conditions. The simple spectrum that remains after subtraction of the reactant ions indicates that m/q 94 and 104 are protonated molecular ions. Thus the EI spectrum contains two compounds with molecular weights of 93 and 103, respectively. Examination of the EI spectrum with knowledge of the molecular weights led to an identification of the two compounds as aniline and benzonitrile. Identification of these two components was confirmed by comparison of GC retention times and GC-MS analysis with those from a mixture of these two compounds [6].

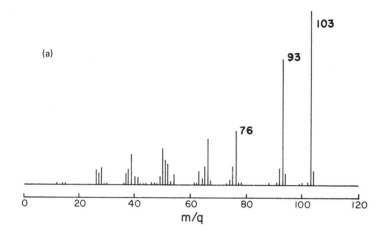

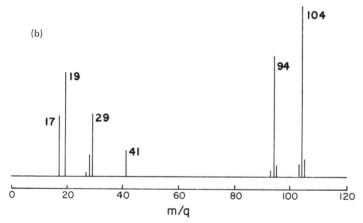

FIG. 4 Mass spectra from compound eluting from gas chromatograph: (a) electron impact; and (b) methane–chemical ionization spectra.

III. APPLICATIONS

In this section, we discuss the applications of EI and CI mass spectrometry to three specific problems. The study of the pyrolysis products from Douglas fir will demonstrate the utility of monitoring specific ions in the mass spectrum. The second example will discuss the analysis of polyurethane degradation products by EI and CI techniques. The last example will describe a technique for the mechanistic study of the thermal degradation products of polyvinyl chloride by effluent gas analysis.

A. Volatile Products from Douglas Fir

Wood is composed of two distinct and chemically different fractions: α-cellulose and lignin. In Douglas fir they occur in approximately a 4:1 ratio. Samples of Douglas fir (0.1 mg) were flash heated to 750°C directly in the injection port of a gas chromatograph and the pyrolysates were separated

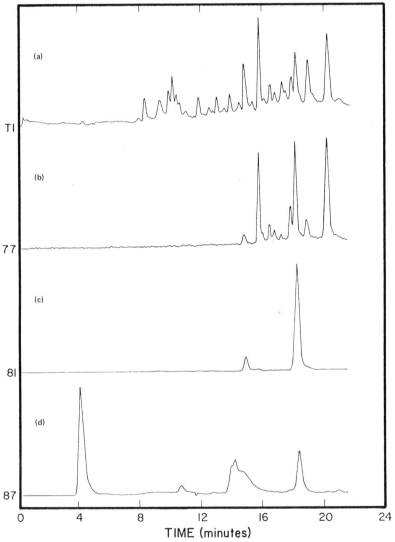

FIG. 5 Selected ion monitoring of the pyrolysis products of wood: (a) total ion reconstruct; (b) m/q 77; (c) m/q 81, and (d) m/q 87.

by a 2 m long by 2.1 mm ID column containing 10% carbowax 20M that was temperature-programmed from 60–220°C at 8°C/min. EI spectra were recorded every 3.5 sec and the total ion plot (summed ion intensity of all ions between m/q 10 and 400) shown in Fig. 5a was produced. Figure 5b is the corresponding selected ion monitor (SIM) plot of m/q 77, $C_6H_5^+$, which is a characteristic fragment for aromatic molecules and therefore an indicator of which GC peaks are likely to contain aromatic species. Lignin, which has a complex chemical structure containing substituted methoxyphenol linkages, is a logical source of aromatic pyrolysates. This observation was supported in other experiments by pyrolyzing α-cellulose under identical conditions and observing that the m/q 77 species was not detected. Thus, the GC peaks in Fig. 5b are derived from the lignin fraction of Douglas fir.

Examination of the EI spectrum from the total ionization peak at 18.4 min did not lend itself to a straightforward interpretation. The mass spectrum recorded at the peak centroid contained intense ions at m/q 77, 81, and 87. This peak was examined further using the SIM technique by plotting m/q 81 and 87 (Figs. 5c and 5d, respectively). The centroid of the three SIM peaks do not coincide, but m/q 77 elutes first, then m/q 81 reaches a maximum intensity, and m/q 87 is on the tailing edge of the peak in Fig. 5a. Thus, the total ionization peak at 18.4 min consists of three or more compounds with slightly different retention times. By carefully manipulating the GC-MS data using computer software, spectra of the two major components were obtained and tentatively identified as furfuryl alcohol ($C_5H_6O_2$) and an isomer of methoxyacetophenone ($C_8H_{10}O_2$). The third constituent could not be identified from the EI spectra.

B. Rigid Polyurethane Foams

A considerable amount of research has been conducted in this and other laboratories on the thermal degradation of flexible and rigid polyurethane foams [7–10]. The initial phase of the degradation involves the cleavage of urethane linkages and formation of volatile high molecular weight polyol fragments. The EI-MS analysis of these products has been hampered by extensive fragmentation of the polyol-type components which show, at best, very low-intensity molecular ions. Figure 6a is an EI spectrum of one isolated pyrolysate fraction from a rigid urethane foam containing polymethylene polyphenylisocyanate (PAPI) and propoxylated trimethylolpropane (TMP). This mass spectrum, which is typical of most polyols, contains severe fragmentation and no detectable molecular ion. The methane–CI spectrum of the same pyrolysate is shown in Fig. 6b. The intense ion at m/q 251 was identified as the M + 1 ion and m/q 279 as the M + 29 adduct [7]. The only fragment ion observed in the CI spectrum was m/q 233, which corresponds to the loss

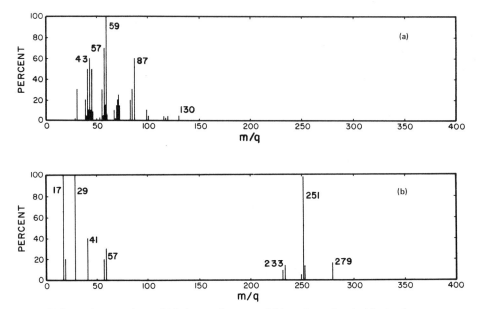

FIG. 6   A comparison of (a) electron impact and (b) methane chemical ionization spectra of dipropoxytrimethylolpropane.

of water from the protonated molecular ion. From this spectrum, the compound was identified as dipropoxytrimethylolpropane. Other fractions of the pyrolysate from this rigid urethane foam gave EI spectra identical to the one shown in Fig. 6a, but the methane–CI spectra were used to unambiguously assign molecular weights of 308, 366, and 424 to higher molecular weight polyols. The CI spectra of these polyols were similar to the one shown in Fig. 6b in that the M + 1 moiety and the loss of one water molecule from its corresponding protonated molecular ion were the only ions that appeared. The use of isobutane as the reactant gas minimized this decomposition pathway and produced very simple spectra containing only the protonated molecular ions of each polyol.

Knowledge of the original polymeric structure is valuable in predicting the types of compounds that may be encountered in polymer degradation studies. Sometimes urethane foam formulations of unknown composition are submitted to this laboratory for pyrolysis studies. A rapid method has been developed for the analysis of the polyol and the isocyanate fractions of rigid urethane foams using CI-MS techniques [11]. Samples of the foam are hydrolysed in aqueous base to the corresponding polyol and polyamine and the hydrolysate is analyzed without further purification. The hydrolysate was fractionated by slowly heating the direct insertion probe from ambient to

325°C over a 10-minute period. The EI spectrum of the polyol fraction [12], shown in Fig. 7a, contains two intense ions at m/q 132 and 146 and several ions of lower intensity at lower masses. The methane–CI spectrum of the polyol, shown in Fig. 7b, contains an intense ion at m/q 279 (M + 1), and low-intensity ions at m/q 307 (M + 29), and 319 (M + 41). The ion at m/q 261 is most likely the loss of water from the protonated molecular ion.

Additional confirmation of the molecular weight of this unknown polyol was provided by the ammonia–CI spectrum of the foam hydrolysates. Polyfunctional alcohols have been reported to lose one or more water molecules from the protonated molecular ion under methane–CI conditions [11]. Using ammonia as the reactant gas, alcohols undergo an ion-molecule reaction in which one H_2O molecule is displaced by NH_3 [13]:

$$ROH + NH_4^+ \longrightarrow RNH_3^+ + H_2O$$

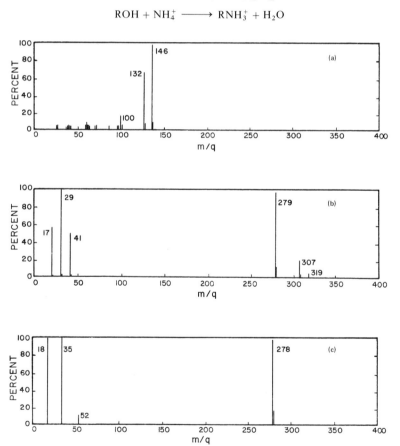

FIG. 7   A comparison of (a) electron impact, (b) methane chemical ionization, and (c) ammonia chemical ionization mass spectra of an unknown polyol.

Figure 7c contains the ammonia–CI spectrum of the polyol fraction from the unknown foam formulation. The spectrum is free of all interfering ions (m/q 18, 35, and 52 are NH_4^+, $(NH_3)_2H^+$, $(NH_3)_3H^+$, respectively) and contains only a pseudomolecular ion M^+ at m/q 278, confirming the molecular weight of the polyol. Comparison of the EI and CI spectra with spectra from known polyol formulations led to the assignment of the following structure:

$$
\begin{array}{c}
\text{HOCH}_2\text{CH}_2 \diagdown \qquad\qquad\qquad \overset{\displaystyle \text{OH}}{\diagup\text{CH}_2\text{CHCH}_3} \\
\text{N—CH}_2\text{CH}_2\text{—N} \\
\text{CH}_3\text{CHCH}_2 \diagup \qquad\qquad\qquad \diagdown\text{CH}_2\text{CHCH}_3 \\
\underset{\displaystyle \text{OH}}{} \qquad\qquad\qquad\qquad \underset{\displaystyle \text{OH}}{}
\end{array}
$$

This structure was consistent with data from other chemical tests used to characterize the polyol.

C. On-Line MS Analysis of Polyvinyl Chloride Pyrolysis

The crucial information in kinetic and mechanistic studies of polymer degradation in the experiments described thus far is the time (and temperature) dependence of product formation. Time sequence information has been generated in this laboratory by coupling the mass spectrometer directly to the pyrolysis unit via a molecular leak valve and heating the sample at a rate of 100°C/min up to a specific temperature. The molecular leak is required since the thermoanalyzer furnace, which provides pressure control of the pyrolysis process, normally operates at atmospheric pressure, while the mass spectrometer is operated at a vacuum of 4×10^{-5}–6×10^{-6} torr.

Polyvinyl chloride was chosen for preliminary kinetic studies because there is a voluminous amount of data in the literature with which to compare experimental results [13]. The transfer lines between the thermoanalyzer, the molecular leak valve, and the mass spectrometer were Teflon coated and heated to 180°C. This was done to reduce adsorption of HCl by metallic surfaces. To calibrate the system and demonstrate performance, HCl and benzene were injected into the furnace with helium flowing through the system at a rate of 30 ml/min. The response to an injection of HCl is shown in Fig. 8. Both major products of PVC pyrolysis, HCl and benzene, were shown to reach the mass spectrometer in 4 sec, and the clearing time (to less than 1% of the ion current from m/q 36) was less than 30 sec. A computer program has been written to assess the diffusion convolution by separating thermal degradation kinetics from mixing and diffusion effects of compounds eluting from the furnace. Application of this program using a linear convolution model shows that the introduction of a compound as an impulse function will reach the mass spectrometer with only minor convolution effects.

The result of preliminary work in which PVC was decomposed at 240°C is shown in Fig. 9. The specific ion abundance of m/q 36 and 78 for HCl and

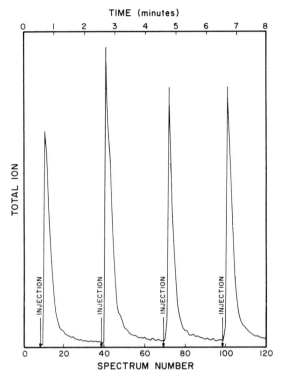

FIG. 8   Detection of m/q 36 at mass spectrometer after repetitive injections of HCl at furnace.

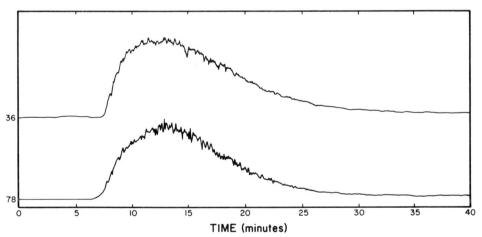

FIG. 9   Single ion monitoring of HCl (m/q 36) and benzene (m/q 78) production from pyrolysis of PVC at 240°C.

benzene, respectively, were monitored continuously during the experiment and the evolution of these materials started simultaneously after 7 min, which corresponds to the point where the thermogravimetric analysis (TGA) indicated that the sample started to lose weight rapidly. Throughout the monitoring, HCl and benzene evolved at a uniform ratio of 70:1.

Another experiment involving specific ion monitoring is illustrated in Fig. 10. PVC was decomposed at a heating rate of 10°C/min in a helium atmosphere and the total ion current trace and four selected ions were monitored. The first gases to evolve were alkenes (up to C_5 and C_6), having m/q 41 as a characteristic fragment ion. The generation of these products corresponds to the first weight-loss process observed in the TGA of the polymer. The next gases that evolved were HCl and benzene (m/q 36 and 78), and these formations were correlated to the second and third decomposition regions of the TGA. At approximately 360°C, a small amount of m/q 91 ($C_7H_7^+$, characteristic of toluene and other alkyl-substituted aromatic compounds) was produced. Evolved gas analysis did not detect any additional degradation products up to 630°C. The continuing weight loss of the PVC

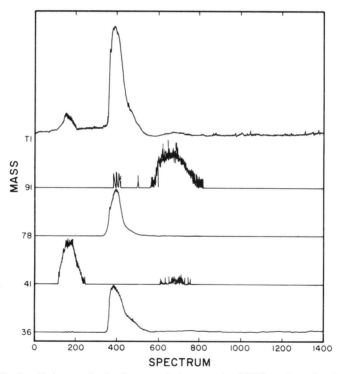

FIG. 10   Specific ion monitoring from pyrolysis as sample of PVC was heated at 10°C/min.

sample between 400–630°C involves the formation of higher weight compounds that are loosely classified as "aerosols."

Analysis of the pyrolysis data for PVC that was heated at 10°C/min was made using the kinetic approach by Mickelson and Einhorn [14]. Figure 11 is a plot of the relative abundance of benzene formation versus time after the sample had been heated to 220°C. The experimental data has been plotted with the theoretical ratio that assumes a 3/2 reaction order. The discrepancy in the fit of the data at low temperature is probably due to degradation that began before a temperature of 220°C had been attained or due to a change in reaction kinetics. Although the results from these kinetic analyses are preliminary, it has been sufficiently complete to show that the combined method of TGA/MS can be successfully applied to the study of pyrolysis with a degree of sophistication never before attained.

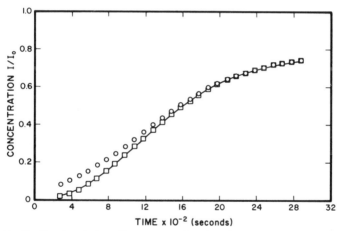

FIG. 11   Predicted evolution of benzene as a function of time using a power model kinetic equation. The temperature is 220°C. ○, predicted concentration with the reaction order equal to 3/2; □, actual data.

IV. CONCLUSIONS

The several techniques described and illustrated in this paper demonstrate that mass spectrometry is a uniquely valuable tool for studying polymer degradation processes. The combination of GC with EI- and CI-MS provides sufficient information to identify most volatile organic compounds encountered in thermal studies. In favorable cases, effluent gas analysis of products can provide kinetic and mechanistic information that is difficult to obtain by any other analytical technique.

ACKNOWLEDGMENTS

The authors wish to acknowledge support by the National Science Foundation under Grants GI-33650 and 6822 and Grant 8244 by the National Aeronautics and Space Administration. Additional support was provided by Hercules Incorporated under Contract PO 0270-04038. We would also like to thank T. A. Elwood, W. H. McClennen, and L. H. Wojcik for their many contributions to this project.

REFERENCES

1. K. J. Voorhees, I. N. Einhorn, F. D. Hileman, and L. H. Wojcik, *J. Polym. Sci., Part B* **13**, 293–297 (1975).
2. C. A. McDowell, "Mass Spectrometry." McGraw–Hill, New York, 1963.
3. S. R. Heller, *Anal. Chem.* **44**, 1951–57 (1972).
4. F. H. Field, *Accounts Chem. Res.* **1**, 42–49 (1968).
5. F. D. Hileman, T. A. Elwood, M. L. Vestal, and J. H. Futrell, paper X-2 presented at *22nd Annu. Conf. Mass Spectrom. and Allied Topics, May 19, 1974.*
6. I. N. Einhorn, D. A. Chatfield, K. J. Voorhees, F. D. Hileman, R. W. Mickelson, S. C. Israel, J. H. Futrell, and P. W. Ryan, *Fire Res.* **1**, 41–56 (1977).
7. F. D. Hileman, K. J. Voorhees, L. H. Wojcik, M. M. Birky, P. W. Ryan, and I. N. Einhorn, *J. Polymer Sci.* **13**, 571–584 (1975).
8. I. N. Einhorn, M. S. Ramakrishnan, and R. W. Mickelson, *Polym. Sci. Engr.* (in press).
9. W. D. Woolley, *Brit. Polym. J.* **4**, 27–43 (1972).
10. N. A. Mumford, D. A. Chatfield, and I. N. Einhorn, *J. Appl. Polymer Sci.* (in press).
11. N. A. Mumford, D. A. Chatfield, and I. N. Einhorn, *Fire Res.* **1**, 107–17 (1978).
12. A. M. Hogg and T. L. Wagabhushan, *Tetrahedron Lett.* **47**, 4827–31 (1972).
13. W. H. Starnes, Jr., *Polym. Prepr., Amer. Chem. Soc., Div. Polym. Chem.* **18**, 493–498 (1977). [Also, see the references therein.]
14. R. W. Mickelson and I. N. Einhorn, *Thermochemica Acta* **1**, 147–153 (1970).

16

Mass Spectrometry of Thermally Treated Polymers

G. J. MOL, R. J. GRITTER, and G. E. ADAMS

IBM CORPORATION
SAN JOSE, CALIFORNIA

I. INTRODUCTION

Any characterization of polymers by mass spectrometry necessarily involves rather complex physical and chemical changes. These changes are brought about by heating a liquid or solid polymer so that it is transformed into a vapor by either a process called thermolysis or one called pyrolysis. In thermolysis, the polymer is heated from room ambient to 200–300°C and then often held isothermally in order to drive off volatiles. Pyrolysis, on the other hand, is where the sample is heated to 300–900°C to cause complete polymer fragmentation. The polymer heating can be ballistic or programmed and varies in rate such that it may be as rapid as 100°C/msec,

257

or as slow as 1°C/min. The heating may take place under vacuum or atmospheric pressure.

One of the most useful techniques has been to combine these thermal treatments with a gas chromatograph because of its ability to separate, detect, and provide a means of identification of small amounts of a wide variety of materials. The inlet port of the gas chromatograph can be equipped with a pyrolysis unit and the characteristic chromatogram (pyrogram) of a polymer, often called a "fingerprint," is obtained. This chromatogram is usually characteristic only for a particular thermal treatment, but is completely reproducible.

The direct identification of the thermolysis/pyrolysis fractions is readily accomplished by the attachment of a rapid scan mass spectrometer to the exit port of the gas chromatograph, so that the effluent of the gas chromatograph passes directly into the mass spectrometer. The mass spectrometer itself provides another thermolysis/pyrolysis method for polymer analysis by use of a direct insertion probe with controlled heating (up to 1000°C) or by heating a portion of a glass inlet system of the mass spectrometer in a controlled manner to volatilize the absorbates into the ion source of the mass spectrometer.

Another very interesting analytic technique has been the heating of polymers in a thermogravimetric analyzer with a directly attached mass spectrometer. Thermogravimetry of a polymer is accomplished as well as identification of residual gasses, solvents, or monomers that are driven off.

Figure 1 shows the general system that is used to examine polymers by these thermal methods. The various combinations of these methods are listed in this same figure.

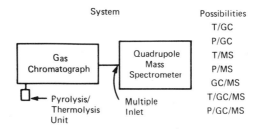

FIG. 1   A typical system and potential combinations.

This paper will indicate how certain of these combinations have been used to provide information on polymer identity, polymer equivalency, polymer structure, and polymer impurities.

II. INSTRUMENTATION AND EXPERIMENTAL TECHNIQUES

The techniques that have been utilized involve several instruments used either alone or in a coupled analytic chain. Four techniques were used:

1. Pyrolysis of the polymer with a pyrolyzer coil or ribbon in the injection port of a gas chromatograph followed by mass spectrometry analysis (pyrolysis/gas chromatography/mass spectrometry—P/GC/MS).

2. Pyrolysis of the polymer in the ion source of a mass spectrometer for mass analysis of the thermal fragments (pyrolysis/mass spectrometry—P/MS).

3. Heating the polymer sample in the ion source of a mass spectrometer to drive out residual volatile materials for mass analysis (thermolysis/mass spectrometry—T/MS).

4. Heating the polymer in the sample holder of a vacuum thermobalance either to drive off residual volatiles or to pyrolyze it (thermogravimetry/mass spectrometry—TG/MS).

For pyrolysis/gas chromatography/mass spectrometry, polymers (50 μg) were generally cast from a solvent solution onto a platinum ribbon of a Chemical Data Systems Pyroprobe 120. The probe was heated in 8–10 milliseconds to 850°C in the injection port of a Perkin–Elmer 3920 gas chromatograph. The effluents were concentrated by a membrane separator and fed into a Hewlett–Packard 5981A mass spectrometer. The mass spectrometer functions as both a gas chromatograph detector and a peak identifier.

The pyrolysis/mass spectrometry technique used the direct insertion probe on the mass spectrometer or a modified platinum coil probe with the Chemical Data System Pyroprobe 120 controller. Less than 100 ng of polymer sample was placed into the nest of the probe. The probe was inserted into the ion source of the mass spectrometer and heated to 600°C. The pyrolysis products were identified by mass spectrometry.

The third technique (thermolysis/mass spectrometry) was identical to the second, with the exception that lower temperatures were used. This technique was used to examine residual monomer, processing solvents, thermal stability of the polymer, and total outgassing (using the total ion monitor of the mass spectrometer). This was also carried out using a heated glass bulb on the inlet system of a Hitachi–Perkin–Elmer RMS-4 mass spectrometer, thus permitting larger samples to be used.

The fourth technique (thermogravimetry/mass spectrometry) is a variation of T/MS and includes the additional feature of a Mettler Thermoanalyzer I. The sample (1–2 mg) was placed in the analyzer pan and heated at 6°C/min. Residual solvents and monomers that were driven off were analyzed by a UTI 100C Precision Mass Analyzer, which was coupled to the thermoanalyzer by attaching an open connector to the quartz furnace tube. The direct coupling to the furnace tube eliminates the problems inherent in trapping or collecting fractions for subsequent analysis, such as secondary reactions and/or contamination of products from more than one reaction.

The schematic of the Mettler Thermoanalyzer I coupled to the UTI 100C Precision Mass Analyzer (F) is shown in Fig. 2.

A Cajon "flexible glass-end tubing" made of 321 stainless steel fused to a type 7740 Pyrex glass on one end served as the interface between the UTI 100C flange and the quartz furnace tube of the thermoanalyzer. A graded quartz-to-Pyrex seal was used for the furnace connection. Quartz tabs were made on the furnace to hold the Nichrome heating element below the interconnecting tube. The 321 stainless steel was vacuum welded to a $2\frac{3}{4}$ inch "Con Flat" flange.

The distance between the sample pan (R) (4-mm high and 3-mm diameter) and the quadrupole ion source is 20 cm. The interconnecting orifice has a 23-mm inner diameter, which, along with its short length, results in a negligible dead time. The ability of the quadrupole mass filter to accept the ions over a relatively large inlet-energy spread and at varying entrance angles makes it well suited for use in this application.

A clamping and pneumatic lifting mechanism supports the mass analyzer and facilitates sample changes by eliminating the need to break the furnace-to-mass analyzer seal. The pneumatic cylinder is a dual action type with rate controllers on each air line. The power supply for the mass analyzer is

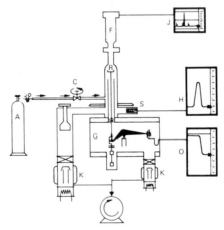

FIG. 2   Instrumentation. A, C, gas inlet system; F, mass spectrometer; G, balance chamber; H, pressure record; J, mass record; K, diffusion pumps; O, balance record; R, reaction/sample chamber; S, ionization gauge.

mounted on a stand attached to a tube inserted into the high-temperature furnace support tubing. The quadrupole electronic control console is mounted in the Mettler electronic cabinet.

Signals from the QMS multiplier after amplification are displayed on an oscilloscope or a two pen recorder (J) at slow (2-min) scans. One 60 and a 250 sec^{-1} diffusion pump (K) evacuate the balance housing (G) and sample chamber. Total pressure is determined with an ionization gauge (S) and is continually recorded (H) during the analysis. Pressures as low as 2×10^{-6} torr have been achieved with this system. The TG, DTG, DTA, pressure and temperature signals are recorded (O) on a multipoint recorder. Purge gas is furnished from cylinder (A) through metering valve (C).

III. RESULTS AND DISCUSSION

A. Analysis of Chemical Equivalency

It is often important to show the equivalency of polymers from different manufacturers for either multiple or alternate sourcing. The latter was the case when the manufacture of a glass-filled polycarbonate was stopped. The polycarbonate from another manufacturer had comparable mechanical and physical properties; however, chemical equivalency was also needed. Pyrolysis gas chromatography was used to show chemical equivalency. The pyrograms are shown in Fig. 3.

The method of polycarbonate synthesis determines the nature of the degradation products. Thus, these pyrograms provide information on the method of synthesis. The decomposition product, *p-tert*-butyl phenol (the polymer end group), is obtained from solvent synthesized polycarbonate. In the solvent method,

is the end group produced by phosgene, the sodium salt of Bisphenol A and and *p-tert*-butyl phenol.

The end groups,

produced by Bisphenol A and diphenyl carbonate, which are used in the melt method, yield phenol and Bisphenol A on decomposition.

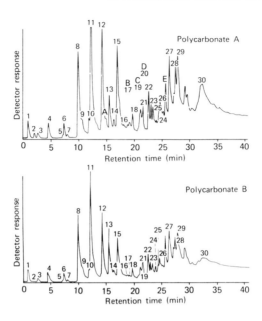

Peak No.	Compounds	Peak No.	Compounds
1	CO₂	19	Diphenylmethane
2	CHCl₃ (solvent peak)	20	Phenyl p-tolyl ether
3	Benzene	21	2-Ethylbiphenyl
4	Toluene	22	Fluorene
5	Ethylbenzene	23	Di-p-tolyl ether
6	Xylene	24	Diphenyl carbonate
7	Styrene	25	2-Phenyl-2-phenylpropane
8	Phenol	26	7-Phenyl benzofurane
9	Methylstyrene	27	p-α-Cumylphenol
10	o-Cresol and indene	28	Butyl phthalate
11	p-Cresol	29	Biphenyl tert-butyl ether
12	p-Ethylphenol	30	Bisphenol A
13	Propylphenol		
14	Isopropylphenol	A	Bromocresol
15	Isopropenylphenol and p-tert-Butylphenol	B	Dibromotoluene
16	Diphenyl ether	C	Ionol and dibromo compound
17	Diphenyl ether	D	Dibromo compound
18	2-Methylbiphenyl	E	γ-Bromopropylbenzene

FIG. 3   Pyrolysis/gas chromatography/mass spectrometry of polycarbonates.

Both polycarbonates have a large *p-tert*-butylphenol peak (peaks 15 in Fig. 3) characteristic of polycarbonate synthesized by the solvent method. This compound is not found in the pyrogram of melt synthesized polycarbonates [1]. The remainder of the pyrogram would be very much the same, regardless of the method of synthesis.

The molecular weight can be determined by the yield of the end-group *p-tert*-butyl phenol, which decreases as a function of molecular weight. From the table used by Shin Tsuge *et al.* [1] a yield of approximately 5% would correspond to a molecular weight of 24,000.

The correspondence of identified peaks in the two pyrograms is good, both having the same polymer pyrolysis fragments as well as the same butyl phthalate plasticizer. Also, both have brominated fire retardants which decompose in pyrolysis to produce bromophenol fragments. Polycarbonate A had a much higher concentration of fire retardant, which made its detection and identification easier. Ionol, an antioxidant, was only found in polycarbonate A. However, all characteristics considered, both polymers are essentially equivalent.

Two du Pont Teflon polymers, tetrafluoroethylene (TFE) and fluorinated ethylene propylene (FEP), are frequently used interchangeably, although

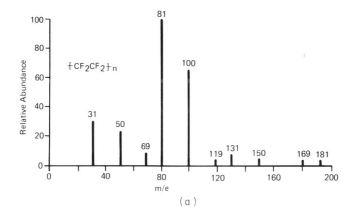

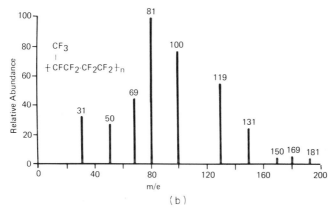

FIG. 4 Pyrolysis/mass spectrometry of Teflon polymers. (a) Teflon–TFE; (b) Teflon–FEP.

their properties differ. TFE has greater fatigue resistance, higher thermal stability, lower coefficient of friction, and better chemical stability, while FEP is less permeable and has better abrasion resistance, lower cold flow properties, and higher impact strength. As shown in Fig. 4, the mass spectrum of the pyrolysis products of FEP has more abundant m/e 69 (CF_3+), m/e 119 (C_2F_5+), and m/e 131 (C_3F_5+) peaks.

The m/e 69 (CF_3+) is usually the base peak in compounds having perfluoromethyl groups. While this is not the case in FEP a definite increase in abundance does occur. Thus the abundance of CF_3+, C_2F_5+, and C_3F_5+ distinguish FEP from TFE.

B. Analysis of Polymer Composition

The amount of fire retardant, triphenyl phosphate, in a commercial polymer blend (polystyrene–polyphenylene oxide) greatly influences the parameters required for injection molding. Excess triphenyl phosphate will bleed out during molding, filling the cored holes in the tooling.

The initial weight loss in the thermogravimetry curve (Fig. 5) was that of the fire retardant as shown by its mass spectrum (Fig. 6). The weight loss of 10 percent thus represents that of the fire retardant.

The composition of the blend was determined by pyrolysis gas chromatography/mass spectrometry. In the pyrogram shown in Fig. 7, the ratio of the area of styrene (peak 4) to dimethyl phenol (peak 9) was used with a

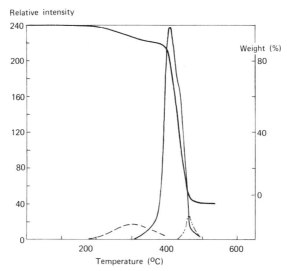

FIG. 5   Thermogravimetry/mass spectrometry of polystyrene–polyphenylene oxide blend. ———, 104; ————, 122; —··—··, 326.

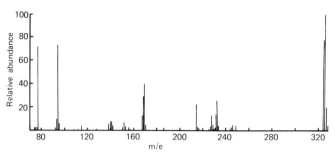

FIG. 6   Mass spectrum of triphenylphosphate obtained during initial weight loss of polymer blend.

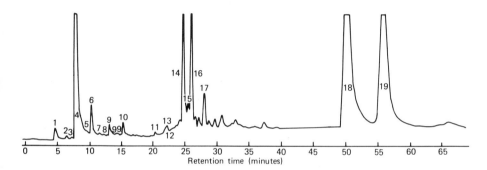

Peak No.	Compounds	Peak No.	Compounds
1	Toluene	11	Diphenylmethane
2	4-Ethenyl 1-cyclohexene	12	1, 1-Diphenylethane + Ionol
3	Ethylbenzene	13	1, 2-Diphenylethane
4	Styrene	14	1, 3-Diphenylpropane
5	β-Methylstyrene	15	Diphenylpentane
6	α-Methylstyrene	16	1, 3-Diphenylbutane
7	o-Methylstyrene	17	$C_{16}H_{16} + C_{16}H_{14}$
8	Cresols	18	Triphenylphosphate
9	Dimethylphenols	19	Diphenylpropane (dimer)
10	Trimethylphenol		

FIG. 7   Pyrogram of polystyrene and polyphenylene oxide blend.

calibration curve to determine the percent composition of styrene and polyphenylene oxide.

The calibration curve was obtained by preparing various solutions of known concentration of styrene and polyphenylene oxide. The percent composition for this sample is 55% styrene, 34% polyphenylene oxide, and (from the TG/MS data) 10% triphenyl phosphate. The ratios for another polyblend were 37:60:3 for polystyrene: polyphenylene oxide: triphenyl phosphate.

The fact that this is a blend is also evident from the maxima in the evolution of styrene and dimethyl phenol which occur at two distinct temperatures (Fig. 5).

FIG. 8   Resist processing.

C. Mass Spectrometry Analysis as an Aid in Resist Material Selection and Processing

Resist processing, whether the resist is a photo, electron, x ray, or any other type of resist, is based on having a polymer or polymer system attain certain specific characteristics. Figure 8 systematically indicates the steps usually taken in resist processing. A good resist depends on the nature of the original polymer, the integrity of the polymer film, the changes occurring during solvent removal, the sensitivity during exposures, and the absence of impurities.

While commercial plastics are known to contain many additives in addition to the base polymer, a newly synthesized polymer is often thought of as being pure. The identification of the most abundant materials outgassing from a newly synthesized polymer (following precipitation and drying) was obtained by heating in the ion source of a mass spectrometer (Fig. 9). Toluene was the synthesis solvent and methanol was the precipitation solvent for the resist polymer of interest. The dioxane was absorbed in the drying oven from another polymer. In another case, ethyl ether, the precipitation solvent, was found to form a stable complex that decomposes at 208°C. A change in the precipitation solvent and a cleaner drying environment produced a polymer which would be expected to form a better film for resist processing.

A coated methacrylate film was examined by thermogravimetry/mass spectrometry. The apparent desorption of water, loss of coating solvent, and depolymerization occurs as the sample is slowly heated (Fig. 10). Because of

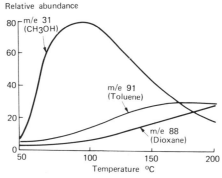

FIG. 9   Thermolysis/mass spectrometry of newly synthesized polymers showing synthesis solvents and dioxane contamination.

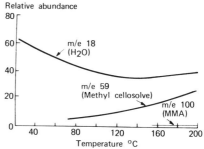

FIG. 10   Thermogravimetry/mass spectrometry of coated polymethacrylate–acrylic acid resist film.

its high boiling point the coating solvent, methyl cellosolve, comes off quite slowly and requires considerable heating. These higher temperatures are obviously deleterious to the polymer as shown by the appearance of methyl methacrylate (MMA). Figure 10 also gives some indication that chemical changes could be occurring during the prebaking operation.

The prebaking of a resist system is usually considered to be just a solvent removal operation. Figure 11 gives evidence that chemical reactions in a methacrylate copolymer can also occur. The data indicate that the water comes off in two stages, which was part of the evidence that there is water desorption (see Fig. 10), but there is also a chemical reaction in the formation of an anhydride from two acid groups in the polymer. The methanol formation and its time factor was especially important, for it clearly showed that

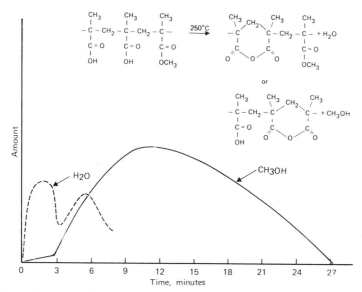

FIG. 11   Thermolysis/mass spectrometry which gives evidence of two different chemical reactions.

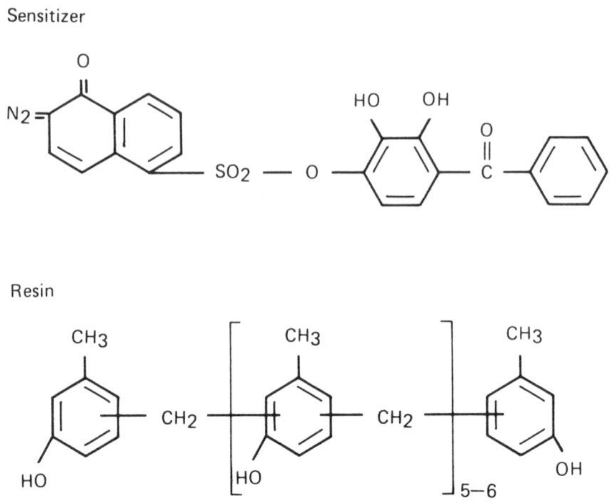

Sensitizer

Resin

FIG. 12   Structural formulas for sensitizer and resin in photoresist.

the anhydride was also forming by the intramolecular reaction of the acid and ester groups in the polymer.

Total outgassing (defined as the area under the total ion curve per unit weight) was used as the selection criteria after prebake for six photoresist films from different manufacturers. Each film was cured according to the recommendations of the respective manufacturer. These resists all have a *m*-cresol–formaldehyde novolak as their base resin and a diazo sensitizer (Fig. 12).

The resists were heated to 240°C, for ten minutes. The mass spectrometer total ion current was monitored as a function of time (Fig. 13). The results are shown in Table I. These outgassing measurements readily indicated potential problem resists, E and F. The photoresist selected was A, based on outgassing and other performance testing.

The mass spectrum of the evolved gases is recorded during the heating cycle. A typical output at 250°C is shown in Fig. 14, with corresponding products identified in Table II.

Pyrolysis of the photoresist was obtained by TG/MS during programmed heating to 650°C.

The SO_2 evolution from the photoresist appears to be Gaussian (Fig. 15), characteristic of a second-order reaction. The phenolic degradation is not a simple or singular reaction and appears to be bimodal in the region of the SO_2 evolution. The SO_2 evolution is linear with respect to a second order reaction (Fig. 16). The activation energy for the evolution of SO_2 from the sensitizer is 30 kcal/mol. The reaction order and activation energy were calculated using the method of Broido [2].

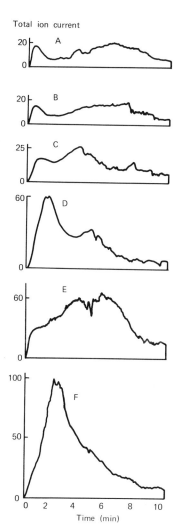

FIG. 13   Outgassing of photoresist per unit weight using total ion monitor.

TABLE I

Outgassing for Photoresist

Sample	Outgassing factor (area/unit wt)
A	12.4
B	19.6
C	25.1
D	36.4
E	53.1
F	62.5

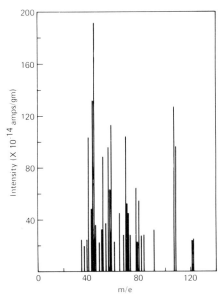

FIG. 14   Mass spectrum of photoresist at 250°C.

TABLE II

Decomposition Products of Photoresist

34	H_2S
41, 55, 69	alkene
43, 57, 71	alkane
44	CO_2
48	SO
64	SO_2
77	ϕ
78	
91	ϕCH_2
107	$HOC_6H_4CH_2$
108	$HOC_6H_4CH_3$
121	$CH_2C_6H_4OCH_3$
122	$HOC_6H_4C_2H_5$

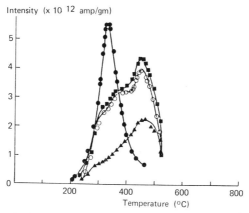

FIG. 15 Thermal degradation of photoresist. ●, SO_2; ■, $\varnothing$; ▲, $\varnothing$-CH_2; ○, $HOC_6H_4CH_2$.

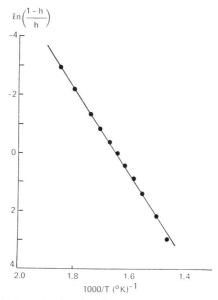

FIG. 16 Second order plot of SO_2 evolution from photoresist.

The SO$_2$ curve was first put into its integral form. The Arrhenius equation used for this curve is

$$\int_h^1 dh/h^2 = (A/u) \int_{T_0}^T e^{-E/RT} dt$$

where u is the heating rate in °C per minute, and h the fraction of SO$_2$ remaining. The final expression for the activation energy obtained from the slope of $\ln((1 - h)/h)$ versus $(1/T)$ is

$$E = S(z)R\Delta \left[\ln\left(\frac{1 - h}{h}\right) \right]/\Delta(1/T)$$

where $S(z)$ is a tabular value.

D. Analysis of Degradation Mechanism

In a different type of study, both capped (formation of a stable hydrazone) and uncapped polylactone polymers were examined to determine whether the mechanism of decomposition was by random chain scission or unzipping. If the degradation mechanism is by random chain scission, then capping the end groups would produce no effect.

The chemical structure of the polylactone of dimethyl ketene and the hydrazone derivative are shown [3] in Fig. 17.

Figure 18 illustrates the thermal decomposition of the uncapped polyester. The evolved gases exhibit the same initial inflection as shown in the weight

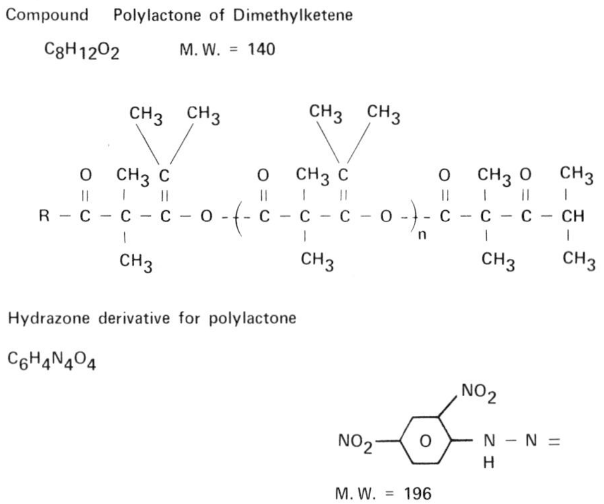

FIG. 17   Chemical structure of the polylactone of dimethylketene and the hydrazone derivative.

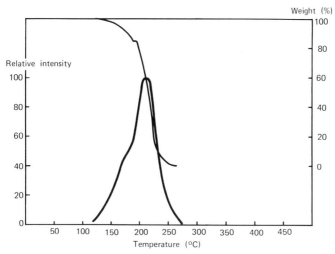

FIG. 18   Thermal degradation of uncapped polylactone. ———, 140 (monomer).

loss curve. The half volatilization point and effluent peak is at 215°C. By
capping this polymer, a pronounced shift from 215°C to 375°C occurs in the
decomposition curve (Fig. 19). The evolution of the hydrazone is concurrent
with that of the monomer (*m/e* 140) and its base peak (*m/e* 81).

Capping only one end of the chain also increases the thermal stability.
However, the uncapped chain end initiates the decomposition followed by

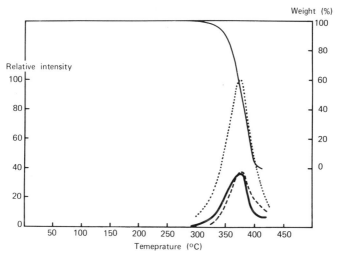

FIG. 19   Thermal degradation of capped polylactone. -----, 81; ———, 140 (monomer);
———, 182 ($C_6H_4N_3O_4$).

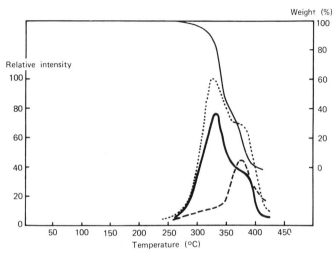

FIG. 20   Thermal degradation of single-end capped polylactone. -----, 81; ———, 140 (monomer); ——, 182 ($C_6H_4N_3O_4$).

the capped end. This is shown by the inflection point in the weight loss curve and the profile of the m/e 182 curve (Fig. 20).

The reason for the positive shift in the degradation of the single capped end is that this sample has a sodium methoxy ketone on the free end instead of the butyl lithium ketone of the previous uncapped polymer. The basic mechanism for the decomposition of this polymer must be unzipping through the ester group. The unzipping mechanism most likely involves a cyclic intermediate of the back biting type [4]. Thus, the thermal stability of this polymer, which degrades by unzipping, may be tailored by selecting the proper end group.

E. Effluent Gas Analysis of Nigrosine Dyes

TG/MS analysis was used in the characterization of three Nigrosine dyes designated as Nigrosine A, B, and C.

A probable structure for the dye is

which is supported by elemental analysis ($C_{18}H_{19}N_3O_7S_1$).

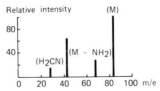

FIG. 21   Cyanoguanidine from Nigrosine A.

Nigrosine A dye (pH = 10) on heating (TG/MS) initially loses cyano-guanidine (Fig. 21), which is probably added as a fixer. At higher temperatures (above 300°C) the dye begins to decompose and the evolution of several characteristic fragments (m/e 64, SO_2; m/e 76, CS_2; m/e 78, benzene; m/e 93, aniline) are shown (Fig. 22).

Similar analysis of Nigrosine B (pH = 5.6) showed no cyanoguanidine and a bimodal distribution in the SO_2 peak due to the presence of both the hydrogen and sodium cations (Fig. 23).

The relative area under each peak is proportional to the amount of each cation. The effect of various cation compositions on the thermal stability and the maxima in the evolution of SO_2 is shown in Table III.

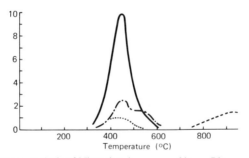

FIG. 22   Effluent gas analysis of Nigrosine A. ———, 64; ---, 76; — ·, 78; · · · · · ·, 93.

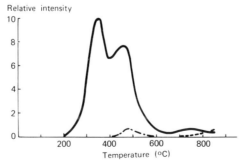

FIG. 23   Effluent gas analysis of Nigrosine B. ———, 64; ---, 76; — ·, 78.

TABLE III

The Effect of Cation Composition on
Thermal Stability of the Maxima in
SO$_2$ Evolution from Nigrosine Dye

Cation	Temperature of maxima for SO$_2$ evolution (°C)
H	340
Na	384
K	444
Ca	444
Mg	510

A third Nigrosine dye had the same pH and sulfur content as Nigrosine B, yet had poorer solubility. In both techniques (TG/MS, T/MS) a new peak at m/e 80 was present. The ratio of this peak to m/e 64, m/e 48, m/e 32 showed this compound to be SO$_3$, the dehydration product of a sulfate or sulfuric acid. The SO$_3$ from sulfuric acid showed up in the T/MS technique (insertion time is in the range of 10–20 seconds) but not in TG/MS, which has a considerably longer pump down cycle. Thus, in TG/MS, only the evidence for a small amount of sulfate is seen (Fig. 24). The free sulfuric acid thus accounted for the poor solubility, due to the lower degree of sulfonation.

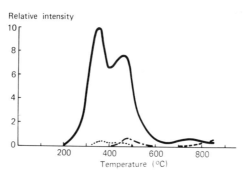

FIG. 24   Effluent gas analysis of Nigrosine C. ———, 64; ---. 76; — ·, 78; · · · · · ·, 80.

IV. SUMMARY

It is obvious that either thermolysis or pyrolysis followed by mass spectrometry or gas chromatography/mass spectrometry can be a very useful method for showing that polymers from different manufacturers may have a similar structure or may be dissimilar although having the same name. In addition, this method can be useful in showing the presence and amount of impurities in polymers and their films. Finally, it is obvious that these techniques can aid in identifying the mechanics of polymer decomposition, and the characterization of various materials.

ACKNOWLEDGMENTS

Most of the synthesized polymers were made by the late Dr. D. E. Johnson, and we must acknowledge his skill and effort in preparing these materials.

REFERENCES

1. S. Tsuge, T. Okumoto, Y. Sugimura, and T. Takeuchi, *J. Chromatogr. Sci.* **7**, 255 (1969).
2. A. Broido, *J. Polym. Sci. Part A-2* **7**, 1761 (1969).
3. E. M. Barrall, II, D. E. Johnson, and B. L. Dawson, *Anal. Calorimetry* **3**, 611 (1974).
4. R. Simha, *Advan. Chem.* **34**, 157 (1962).

AUTHOR INDEX

A

Abragam, A., 57(1), *76*
Achhammer, B. G., 234, *239*
Adams, J. Q., 25, 34, *40*
Albrecht, A. C., 122(1,2), *132*
Alfonso, G. C., 20(4), *39*, 42(14), *54*
Allerhand, A., 58(4), 60(8), 61(15), 62(4), 65(8), 68(8), *77*
Almgren, M., 117(20), *119*
Altamirano, J. L., 156(30), *168*
Amelin, A. V., 223(16,17) 236(16,17), *238*
Anderson, L. R., 2(10), *5*
Andrews, E. H., 222(5), *238*
Asada, T., 186, *204*
Ashby, G. E., 102, *118*
Asher, I. M., 129(12), *133*
Aufdermarsh, C. A., 149(14), *168*

B

Backman, D. K., 223(9), 224(9), 236(9), *238*
Backstrom, H. L. J., 117(18b), *119*
Bailey, H. C., 150(16), *168*
Balaceanu, J. C., 117(19), *119*
Barker, R. E., Jr., 102(5c), *118*
Barnard, D., 83(13), *85*
Barone, B. J., 83(12), *85*
Barrall, E. M., 272(3), *277*
Bascom, W. D., 172(3,4,6), 173, *183, 184*
Baugher, J. F., 223(14), *238*

Baumgartner, W. E., 223(18), *238*
Beavan, S. W., 79(6), 83(6), 84(6), *85,* 116, *119*
Becconsall, J. K., 66(19), 75(19), *77*
Beer, M., 193, *205*
Bennett, J. E., 117(18a), *119*
Berger, P. A., 223(19), *238*
Bernstein, A., 66(20), 75(20), *77*
Bevilacqua, E. M., 79(1,4), 80(1,4), 84(1), *85*
Biagini, E., 42(13), *54*
Birdsall, W. J. M., 72(25), *77*
Birky, M. M., 249(7), *256*
Bloembergen, N., 75(28), *77*
Blossey, E. C., 89(19), 92(19), *99*
Boer, F. P., 180(28), *184*
Boerio, F. J., 174(17), 176(17), 177(17), 178(17), 180(17), *184*
Bovey, F. A., (1), *5,* 32, 34(16), *40*
Boyd, D. B., 127, *132*
Braley, H. B., 173(13), *184*
Brame, E. G., 146(12), 149(12,13), *168*
Breck, A. K., 87(11), *98*
Brey, W. S., 1(2), *5*
Brill, W. F., 83(12), *85*
Broido, A., 268, *277*
Brotskii, A. E., 212(6), *219*
Brown, J. K., 4(17), *5*
Brubaker, M. M., 19(2), 20(2), *39,* 42(16), *54*
Buckhart, R. D., 42(3), 43(3), *54*
Bunn, C. W., 42(15), *54*

279

SUBJECT INDEX

A

Additivity rule for methylene carbons, 14–16
Amino acids, 128

B

Biopolymers, 121
Blend
  polystyrene–polyphenylene oxide, 264–265
  polyvinylidene fluoride–polymethyl methacrylate, 156–161
Bond orientation, 193, 201
Branching, 32–35
Butyl rubber, 61

C

Camphor, 2
Chemical ionization mass spectrometry, 241
Chemical structure, 67
Chemiluminescence, 105–116
Chloroform-d solvent, 21
cis/trans ratio measurement, 9–13
^{13}C NMR, 1, 7, 19, 57, 82, 94–97
^{13}C NMR chemical shift
  acetone, 24
  2-butanone, 24
  2,6-dimethyl-4-heptanone, 24
  2,4-dimethyl-3-pentanone, 24
  2,2-dimethyl-3-pentanone, 24
  2-heptanone, 24
  3-heptanone, 24
  4-heptanone, 24
  2,5-hexanedione, 24
  2-hexanone, 24

3-methyl butanone, 24
5-nonanone, 24
3-pentanone, 24
2,4,4-trimethyl-3-pentanone, 24
3-undecanone, 24
6-undecanone, 24
Coal, 139
Comonomer sequence distribution, 30
Crystallinity, 53

D

Data handling, 227–228
Decoupling, ^{13}C NMR (^{19}F), 2
Deuterochloroform, 8
Dichroic ratio, 202
1,2-dideuterotetrachloroethane solvent, 26
Differential scanning calorimetry, 52
Dipropoxymethylolpropane, 250
Douglas fir, 248
Dual beam mass spectrometry, 245

E

Elastomers
  carbon black filled, 136
  polybutadiene, 137
  polychloroprene, 145–151
Electron impact mass spectrometry, 241
End groups, 32–35
ESR spectroscopy, 209–217
Ethylene–carbon monoxide copolymer, 19, 30, 42
Ethylene–ethyl acetate–carbon monoxide terpolymers, 41–46
Ethylene–ethyl acrylate copolymer, 43–46